Reduced Ice Age Prediction Models

Mack D. Harkins

Abstract

Reduced differential equation (DE) models are often used to understand complex systems. We consider here a class of DE models developed in the scientific literature to explain the interglacial/glacial transitions, also known as "Ice Ages." The different DE models we consider are based on different underlying mechanisms for the process of glacier formation and retreat. Ideally, we would like to determine which of these models does the best job of matching the historical records of the ice ages so that the mechanism for the phenomena can be understood more clearly. Such information could also be used to understand glacier behavior under our current conditions of changing climate and for prediction of future ice ages.

In an earlier work by Crucifix and Rougier (2009), a particle-filter algorithm is used with historical data to determine the parameters for a given DE model and the correspondence of the model predictions with the data. In a subsequent paper (Crucifix (2011)), it is stated that unpublished work following the Crucifix and Rougier (2009) paper found that the particle filter was not able to discriminate between DE models on the basis of their long-term dynamics. Our work here investigates this claim, by implementing a particle-filter method and applying it to several competing DE models for ice ages.

The ice age models we considered are a family of three-state, first-order stochastic dynamical systems for global values of CO_2, glacial ice volume and ocean temperature. In addition to model parameters, the effect of variations in solar energy input from the sun due to long term variations of the Earth's orbit was included as a deterministic forcing term that could be toggled on or toggled off. Some parameters were taken as fixed, while others (typically those associated with nonlinear terms) were taken as unknown and to be determined by the particle filter. The size of the stochastic forcing terms was also unknown, and we considered two cases corresponding to small and large stochastic forcing.

The particle filter method is based on using a large number of discrete samples ("particles") to generate a forward-in-time Bayesian approximation to the probability distributions of the model parameters and model states. As the model and parameter distributions are evolved in time, each data observation is used to select particles which are good predictors of the data and reject particles which are bad predictors of the data. The Liu and West (2001) particle filter algorithm we employ also includes a dispersion term so that the cloud of particles remains dispersed and able to continue to explore parameter space. Computational parameters in the algorithm are the number of samples, the estimated size of measurement error, and a resampling threshold. In our case, we use historical measurements of one of the states in the model (CO_2) to determine the three model states an up to 6 model parameters.

Our results include (i) validation of our numerical model and benchmarking tests of the particle filter on synthetic CO_2 data generated by the numerical model; (ii) application of the particle filter to the actual CO_2 data and comparison of the performance of several DE models; and (iii) application of the particle filter on the actual CO_2 data and comparison of model performance with respect to CO_2 data and ice volume proxy data.

Overall, we find that all of the reasonable models performed comparably with respect to predicting the measured CO_2 data (which was used as input for the particle filter). In the case of predicting the ice volume proxy data (which was not used as an input to the particle filter), we found for each model that including the insolation forcing due to orbital variations was a key factor in getting better predictions of the ice volume, but again found that the different models gave similar levels of performance. Based on our findings, we agree with the assessment in Crucifix (2011) that the particle filter method is not sufficiently sensitive to select between the small number of proposed ice age DE models.

Contents

Chapter 1

Introduction

Reduced differential equation (DE) models are often used to understand complex systems such as the El Niño, which is a climate phenomenon where there is an increase in the sea surface temperatures in the Eastern Tropical Pacific Ocean once every 4 years, with a duration of about a year ([5], p. 4). Other reduced DE models that are more related to what we will study in this dissertation, are the ones which attempt to explain the interglacial/glacial transitions, or, the Ice Ages. There are many different types of reduced models for modelling the Ice Ages, and many give similar qualitative behavior, but each may have different underlying mechanisms for the process of glacier formation and retreat. Ideally, we would like to determine which of these models does the best job of matching the historical records of the ice ages to determine the best model mechanism. Such information could also be used to understand glacier behavior under our current conditions of changing climate and for prediction of future ice ages.

The work in this **book** will explore one approach to modelling the Ice Ages that will treat the DE model and measurements as being stochastic, and we will employ a computational algorithm based on a "particle filter" (which will be defined later) to adaptively determine the likely model parameters and be able to evaluate the quality of the model in comparison to the data. As a validation of our method and codes we compare directly to previous work by Crucifix & Rougier, [4] (from now on be called "Crucifix & Rougier (2009)"). Further validation will be conducted using synthetic data and error analysis in order to quantify how well the "particle filter" is doing on a particular model. The results in the figures were qualitatively similar to the ones in Crucifix & Rougier (2009) – they will not be exactly the same partly due to some differences in the choice of the CO_2 data. Error and sensitivity analysis will be done on several different models using this "particle filter" method to test a claim made by M. Crucifix, which is that "[t]he filter performs poorly on the model selection problem because it fails to discriminate models on the basis of their longterm dynamics" ([3], p. 839). As we'll see later, the results will actually support this claim made by M. Crucifix.

Chapter 2

Background

2.1 Introduction to climate modeling of Ice Ages

An Ice Age is a period of time when the global climate temperature went from warm to cold producing glacial cycles – "[a] cycle consisting of a planet-wide glaciation period and a warm period, as observed repeatedly during the past 800 kyr" ([8], p. 276). This means that the global temperature was so cold that the polar ice extends towards the equator. There are several hypotheses as to why these glacial/interglacial transitions happened. A couple important factors that have been proposed to help explain these transitions are variations in the Earth's orbit (Milankovitch cycles) as well as the greenhouse effect from the CO_2 ([5] p. 272 - 273, p. 293 - 294). Insolation, by definition, is "the amount of solar radiation reaching the Earth by location and by season" ([8], p. 276), and the changes in the insolation due to changes in the Earth's orbit have to do with three main components: (i) eccentricity (how "oval" the orbit is), (ii) obliquity ("the tilt of the Earth's equatorial plane relative to its orbital plane"), and (iii) "precession of the Earth's spin axis around the normal to

the orbital plane", according to Milankovitch ([8], p. 131-132). J. D. Hayes et al. used
Southern Hemisphere ocean-floor sediments dating back 450,000 years to study how the
Earth's orbit might induce the ice ages and they've concluded that the "...climatic variance of
these records is concentrated in three discrete spectral peaks at periods of 23,000, 42,000, and
approximately 100,000 years" ([6], p. 1131). According to Milakovitch theory, these periods
should also appear in the orbital data. Precise numerical calculations of the variations in
the Earth's orbit are now readily available [10]. From the orbital data, we computed the
Fourier power spectra of the orbital parameters and looking at Fig. 2.1(c) we do see peaks
in eccentricity at around 100 kyrs, in Fig. 2.2(c), there are peaks in obliquity around 42
kyrs, and in Fig. 2.3(c), there are peaks in precession index around 23 kyrs, confirming what
Hayes et al. concluded.

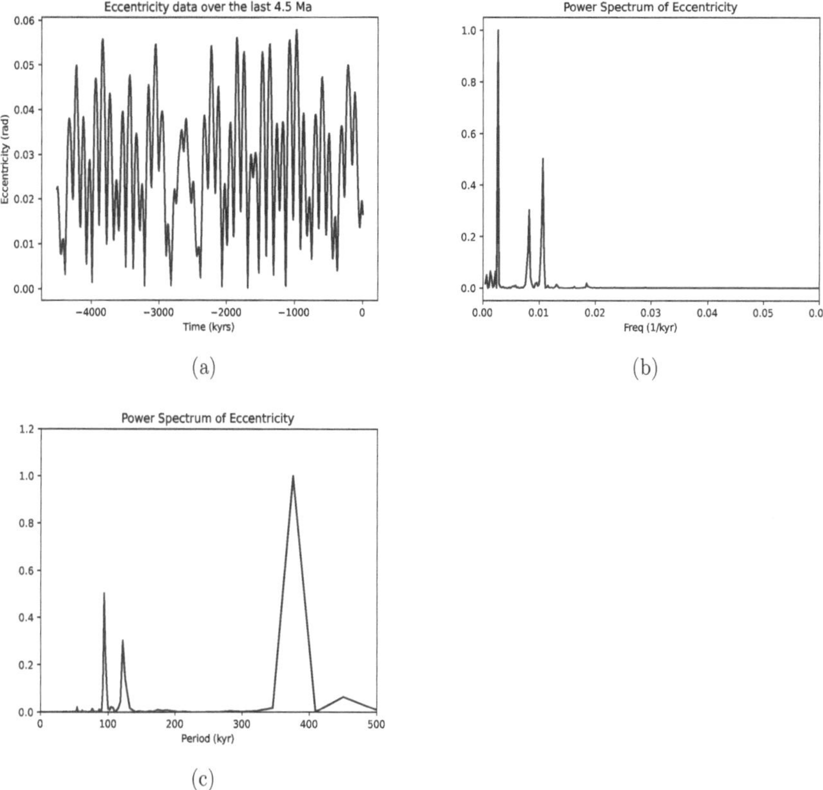

Figure 2.1: Plots of the eccentricity data [10] and its Fourier power spectrum. (a) eccentricity vs time, (b) Fourier power spectrum vs frequency, (c) Fourier power spectrum vs period.

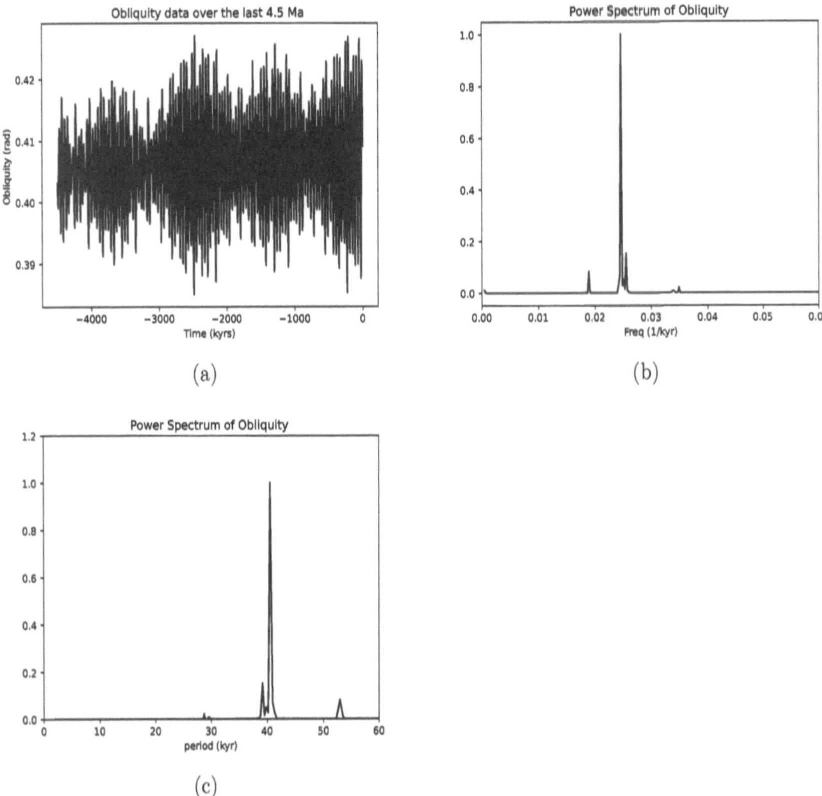

Figure 2.2: Plots of the obliquity data [10] and its Fourier power spectrum. (a) obliquity vs time, (b) Fourier power spectrum vs frequency, (c) Fourier power spectrum vs period.

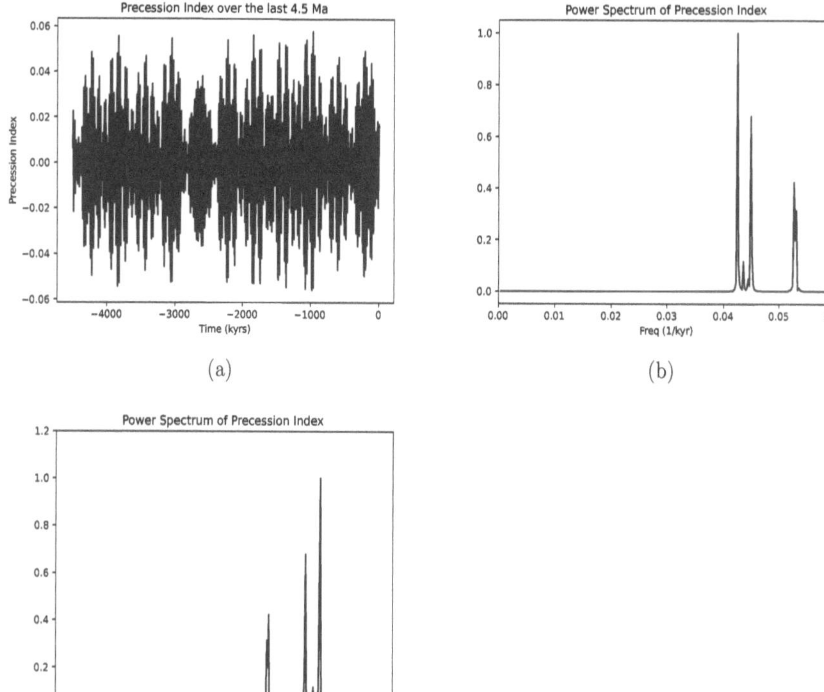

Figure 2.3: Plots of the precession index [10] and its Fourier power spectrum. (a) precession index vs time, (b) Fourier power spectrum vs frequency, (c) Fourier power spectrum vs period.

Indeed both the insolation data implied by the orbital parameters and the $\delta^{18}O$ (a ratio of oxygen isotopes that is used as a proxy for temperature) data, from the Vostok, Antarctica ice core data [17], [16],[18], have some of these peaks in their spectra, as shown in Fig. 2.4(c) and Fig. 2.5(c). However, the 100 kyr peak appearing in the power spectrum of the $\delta^{18}O$ data doesn't even show up in the power spectrum for the insolation data. Interestingly, if we compute the power spectrum of the CO_2 data from the Vostok data set, there's a high peak at around 100 kyrs as depicted in Fig. 2.6(c). The idea is that including the geological carbon cycle as an amplifying effect of the insolation forcing that might help explain the 100 kyr signal in the δ^{18}O data ([5], p. 296). In addition, it would also make sense to include CO_2 in an ice age model because of its role in the greenhouse effect and its subsequent effect on the global temperature: "[a]n increase in the concentration of [greenhouse gases] leads to an increased opacity for infrared radiation of the atmosphere" ([8], p. 276).

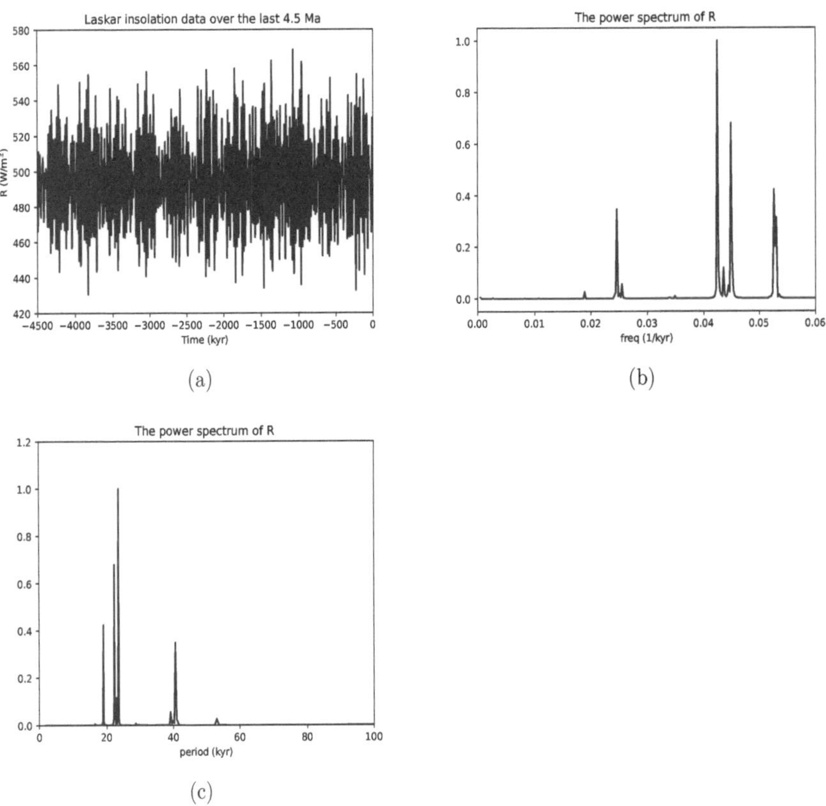

Figure 2.4: Plots of the insolation data [10] and its Fourier power spectrum. (a) insolation data vs time, (b) Fourier power spectrum vs frequency, (c) Fourier power spectrum vs period.

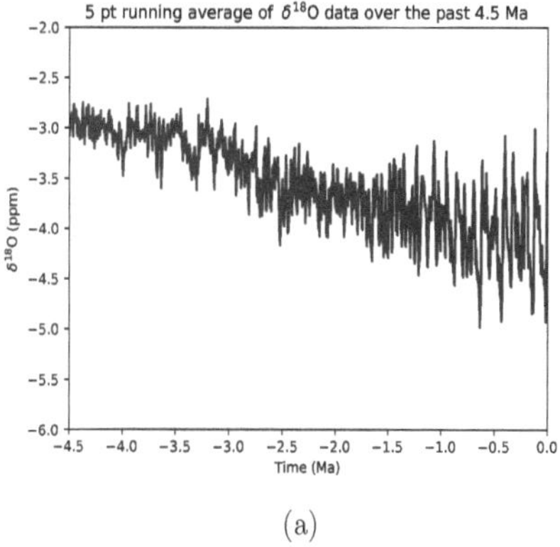
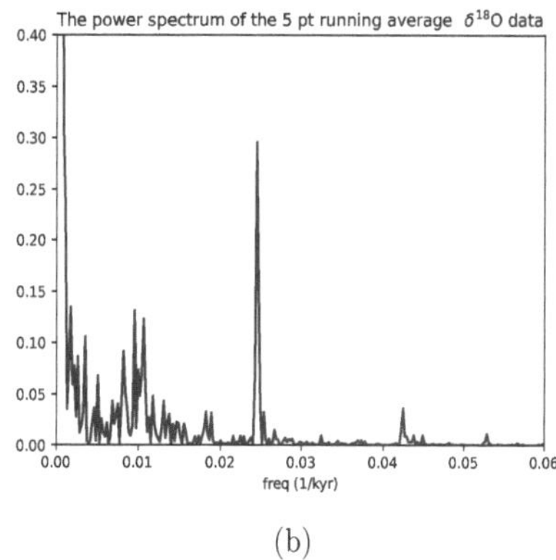
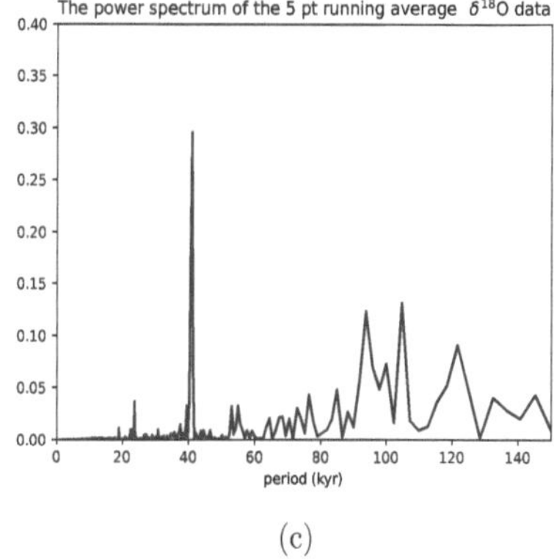

Figure 2.5: Plots of the δ^{18}O data [24] and its Fourier power spectrum. (a) δ^{18}O data vs time, (b) Fourier power spectrum vs frequency, (c) Fourier power spectrum vs period.

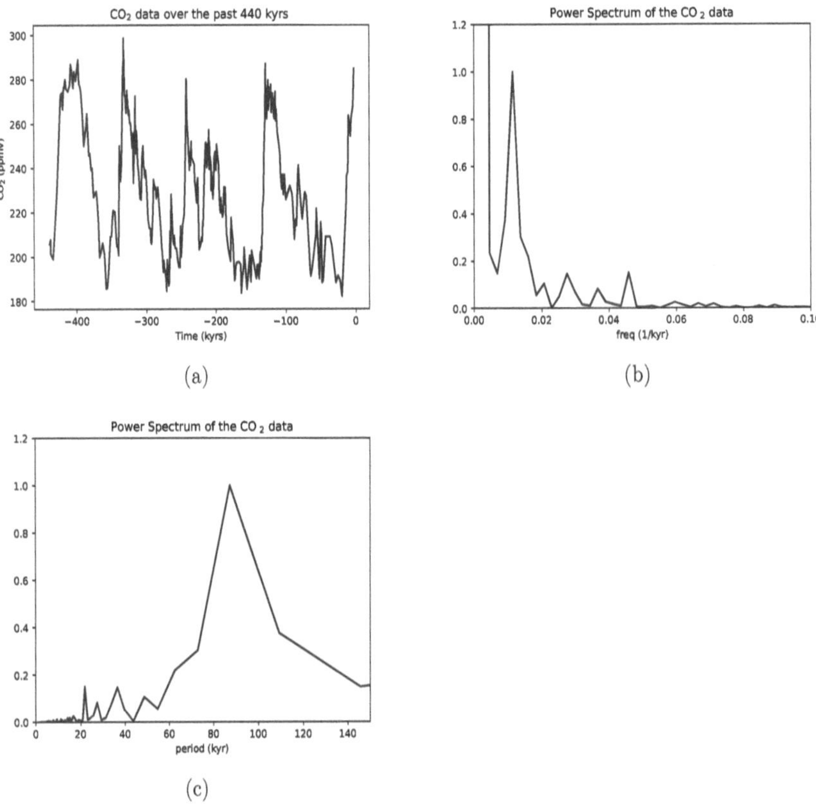

Figure 2.6: Plots of the CO_2 data [17], [16],[18] and its Fourier power spectrum. (a) CO_2 data vs time, (b) Fourier power spectrum vs frequency, (c) Fourier power spectrum vs period.

In this dissertation, we'll study the Pleistocene Ice Age which is an ice age that occurred between 11.7 kyrs to 2.6 Myrs before present [2]. Because we need data dating this far back, we need proxy data to construct what the climate might have been. Data such as $\delta^{18}O$, CO_2 and insolation forcing are used to help paint a picture of what the temperature or air composition was during that time. The CO_2 data is collected from the ice cores, such as from the Vostok, Antarctica ice core data or the European Project for Ice Coring (EPICA) Dome C data [14]. In this paper we'll use the Vostok ice core data from 0 - 440 kyrs before present from [17], [16], [18]). The insolation data is typically calculated numerically from physics-derived equations for the motion of planets [11]. Here, we will generate the 65°N June insolation data from the Fortran programs listed in [10] dating 450 kyrs before present. Some limitations of this data are:

- only a small area of the Earth (65° N) is used as the key forcing in how the glaciers behave; this is traditionally used because it "determines whether snow will be left at the end of the Northern Hemisphere Summer season." [5]

- the proxy data for the CO_2 has gaps in it because it is taken from multiple core samples at different locations and is combined to achieve a "best estimate."

2.2 Introduction to differential equation (DE) modeling of Ice Ages

Applied modeling is sometimes based on simple models in order to help understand the mechanisms underlying the phenomena. Because the complexity of Earth systems (multiple length and time scales) makes it difficult to model from first principles, the following models are constructed ad hoc but incorporate some of the feedbacks and cycles that are thought to be driving factors causing glacial/interglacial transitions. Some of the variables that can influence these transitions are the height-mass balance feedback ([5], p. 277 - p. 281), precipitation-temperature feedback ([5], p. 281 - 283), and carbon cycles ([5], p. 293 - 298). Finally, Milankovitch forcing is also added as the "pacemaker" for the model ([19], p. 201).

The main model of interest is the Saltzman and Maasch 1991 model (SM91):

$$\frac{dI'}{dt} = -a_1\mu' - a_2 I' - a_1 c^{-1}\kappa_\theta \theta' - a_1 c^{-1}\kappa_{\mathscr{R}}\mathscr{R}'(t) + \mathscr{W}_I$$

$$\frac{d\mu'}{dt} = b_1\mu' - b_2\mu'^2 - b_3\mu'^3 - b_4\theta' + \mathscr{W}_\mu \tag{2.1}$$

$$\frac{d\theta'}{dt} = -c_1 I' - c_2\theta' + \mathscr{W}_\theta$$

where I', μ', and θ' are the global ice mass, CO_2 and ocean temperature anomalies, $\mathscr{W}_v$ is the stochastic forcing added to the DE for variable v, $\mathscr{R}'$ is the insolation forcing anomaly, and a_i, b_i, c_i, c, and κ_i are constants ([19], p. 202 - 203). Fig. 2.7 is Fig. 1 taken from [19] on p. 203, showing in detail how the variables in this model are related.

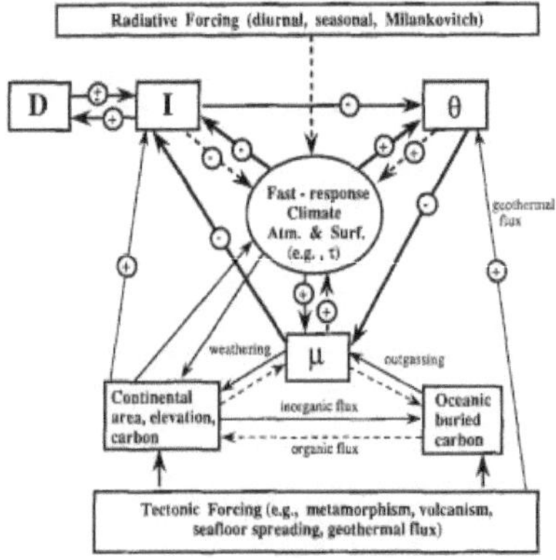

Fig. 1. Schematic diagram showing the interactions implied by the model between the three slow-response variables I, μ, and θ, the fast-response "climate" variables, such as τ (*inner circle*), and external forcing due to insolation changes (*upper box*) and tectonic variations (*lower box*). The effects of bedrock depression, not included in this model, are represented by D

Figure 2.7: Schematic diagram from [19] of how the variables in the SM91 model are related.

Stability and limit cycles of the SM91 model

Here we present a mathematical analysis of the equilibrium points and their stability for the SM91 model. From the analysis we clarify the importance of nonlinear terms in the model in generating limit cycles, and give a characterization of the parameter space that gives rise to limit cycles with periods on order of 100 kyr.

Saltzman & Maasch stability analysis of the SM91 model

The SM91 system was shown to admit free (unforced) oscillations with period on the order of 100 kyr ([19], p. 205). Briefly, the unforced system can be nondimensionalized to obtain

$$\frac{dX}{dT} = -X - Y - vZ$$
$$\frac{dY}{dT} = -pZ + rY - sY^2 - Y^3 \qquad (2.2)$$
$$\frac{dZ}{dT} = -q(X + Z)$$

where $(X(T), Y(T), Z(T))$ are nondimensional versions of $(I'(t), \mu'(t), \theta'(t))$ and p, q, r, s and v are nondimensional parameters. The system always admits the trivial solution $(X, Y, Z) = (0, 0, 0)$, and, because of the cubic term in Y, can admit possibly two more equilibrium solutions (X_1^*, Y_1^*, Z_1^*) and (X_2^*, Y_2^*, Z_2^*) depending on the choice of the 5 parameters. Fig. 2.8 is a reproduction from [19] (Fig. 2) showing the existence and stability of equilibrium solutions in the (r,s) parameter space for fixed values of $p = 1.0$, $q = 2.5$ and $v = 0.2$. From numerical experimentation, Saltzman and Maasch [19] find "that free oscillations indeed tend to exist for all values of r and s not admitting a stable equilibrium that can serve as an attractor." Fig. 2.9 (reproduction of Fig. 3 from [19]) shows the region in (r,s) parameter space that admits free oscillatory solutions along with contours showing the period of oscillation (which are on the order of 100 kyr). While these results are for a specific choice of p, q and v, the results here imply that ~ 100 kyr period oscillations may occur in some contiguous region of the 5-dimensional (p,q,r,s,v) parameter space, and by extension, to a region in the parameter space for the original dimensional model.

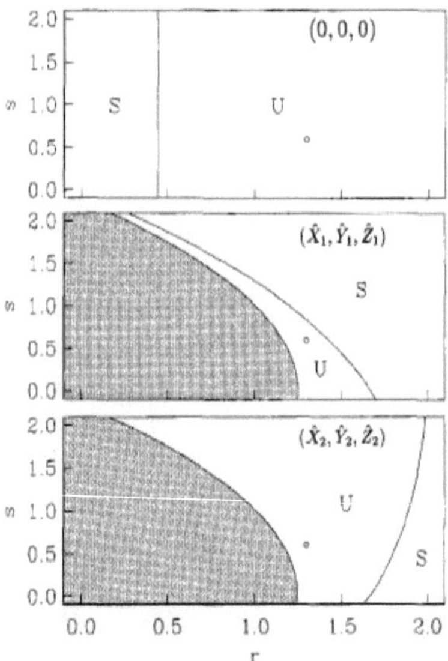

Fig. 2. Stability of three possible equilibrium points of the unforced system (Eqs. 14–16), as a function of the parameters r and s, for assigned values $p=1.0$, $q=2.5$, and $v=0.2$. *Top, middle,* and *lower panels* show the stable (S) and unstable (U) regions corresponding to the equilibrium at $(X, Y, Z)=(0,0,0)$, $(\hat{X}_1, \hat{Y}_1, \hat{Z}_1)$, $(\hat{X}_2, \hat{Y}_2, \hat{Z}_2)$, respectively; the *shaded region* denotes the presence of only a single equilibrium at $(0,0,0)$. The *dot* represents the point in r-s parameter space for the special solution described in the third part of the fourth section

Figure 2.8: The stability region from [19] for equilibrium solutions in the (r,s) parameter space.

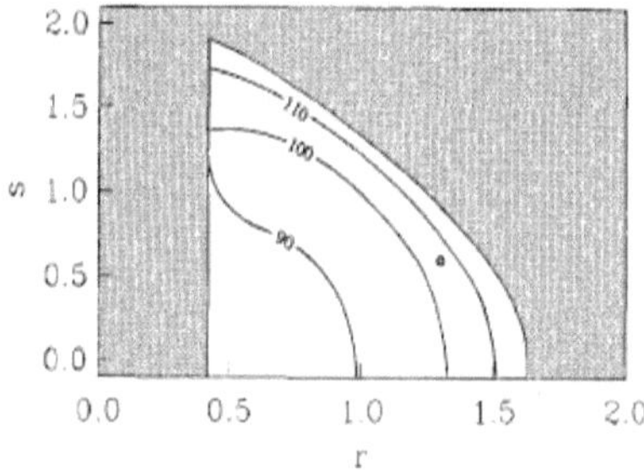

Fig. 3. Region of *r-s* parameter space for which free oscillatory solutions exist. *Isopleths* denote the period of the solution in units of ky. The dot represents the parameter values chosen for our special solution described in the fourth section

Figure 2.9: The region of free oscillatory solutions from [19] in (r,s) parameter space.

Limit cycles and the nonlinear terms in the SM91 model

The SM91 system was shown to admit free (unforced) oscillations with period on the order of 100 kyr ([19], p. 205). We include here a summary of the stability of the SM91 system to clarify the role of nonlinear terms in creating a limit cycle.

First, we nondimensionalize the unforced system (2.1) using time scale $t_c = 1/K_I$, and ice, CO_2, and temperature scales $I_c = a_1 k_\mu b_2/(K_I b_3)$, $\mu_c = b_2/b_3$, and $\theta_c = c_1 a_1 k_\mu b_2/(K_I b_3 K_\theta)$. Letting $T = t/t_c$, $X = I/I_c$, $Y = \mu/\mu_c$, and $Z = \theta/\theta_c$, the nondimensional system becomes

$$\frac{dX}{dT} = -X - Y - \nu Z$$
$$\frac{dY}{dT} = -\rho Z + rY - \varepsilon \left(Y^2 + Y^3\right) \tag{2.3}$$
$$\frac{dZ}{dT} = -q(X + Z)$$

where the nondimensional parameters are

$$\nu = \frac{k_\theta c_1 a_1}{K_\theta K_I}$$

$$\rho = \frac{b_\theta c_1 a_1 k_\mu}{K_\theta K_I^2}$$

$$r = \frac{b_1}{K_I}$$

$$q = \frac{K_\theta}{K_I}$$

In contrast to the choice of scalings in [19] in which the presence of the cubic nonlinear term Y^3 was unaffected by parameters (see Eq. (2.2)), our choice of scalings is chosen so that the size of the nonlinear terms can be controlled by a nondimensional parameter ε. Even so, there is a correspondence between our nondimensional parameters and those of Eq. (2.2) [19] where

$$v = \nu$$

$$p = \rho$$

$$s^2 = \varepsilon$$

and r, q are defined identically in both systems.

Our nonlinear system always has the trivial solution

$$(\tilde{X}_0, \tilde{Y}_0, \tilde{Z}_0) = (0, 0, 0)$$

and has nontrivial solutions given by

$$\tilde{Y}_{1,2} = -\frac{1}{2} \pm \frac{1}{2}\sqrt{1 + \frac{4}{\varepsilon}\left(r - \frac{\rho}{1-\nu}\right)} \tag{2.4}$$

$$\tilde{X}_{1,2} = -\tilde{Y}/(1-\nu) \tag{2.5}$$

$$\tilde{Z}_{1,2} = \tilde{Y}/(1-\nu) \tag{2.6}$$

for

$$\varepsilon \geq -4\left(r - \frac{\rho}{1-\nu}\right).$$

The linear stability of the equilibrium solutions can be found using the Jacobian,

$$J\left(\tilde{X}, \tilde{Y}, \tilde{Z}\right) = \begin{pmatrix} -1 & -1 & -\nu \\ 0 & w & -\rho \\ -q & 0 & -q \end{pmatrix}$$

where

$$w = r - \varepsilon\left(2\tilde{Y} + 3\tilde{Y}^2\right)$$

for some critical point $(X, Y, Z) = (\tilde{X}, \tilde{Y}, \tilde{Z})$.

The eigenvalues λ of J satisfy the cubic equation

$$\lambda^3 + a_2\lambda^2 + a_1\lambda + a_0 = 0$$

where

$$a_2 = 1 + q - w \tag{2.7}$$

$$a_1 = q(1 - \nu) - w(1 + q) \tag{2.8}$$

$$a_0 = q(\rho - w(1 - \nu)) \tag{2.9}$$

Linear stability of an equilibrium solution corresponds to $\text{Re}(\lambda) < 0$. Using the Routh-Hurwitz criterion, $\text{Re}(\lambda) < 0$ if and only if

$$a_2 > 0$$

$$a_1 > 0$$

$$a_0 > 0$$

$$a_2 a_1 > a_0.$$

In general, these inequalities mark out a region in five-dimensional parameter space $(\nu, r, \varepsilon, \rho, q)$ corresponding to stability of equilibrium solutions. For our purposes here, we seek to show how the parameter ε associated with the nonlinear terms in the nondimensional system controls the existence of limit cycles. To this end, as in [19], we consider three of the parameters as fixed ($\nu = 0.2$, $\rho = 1.0$, $q = 2.5$). Based on the findings of Saltzman and Maasch [19], we search for limit cycles using parameters that admit only unstable equilibria. Our results will describe limit cycles (and their periods) in a region of two-dimensional parameter space (r, ε).

Using the fixed parameters (ν, ρ, q) as described above, the condition for existence of the nontrivial solution Eq. (2.3) becomes

$$\varepsilon > -4r + 5$$

Fig. 2.10 shows the range of existence of the three possible equilibrium points. The upper plot shows that the trivial solution exists for all ε and r, while the lower two plots show that the nontrivial solutions exist only for sufficiently large ε and r.

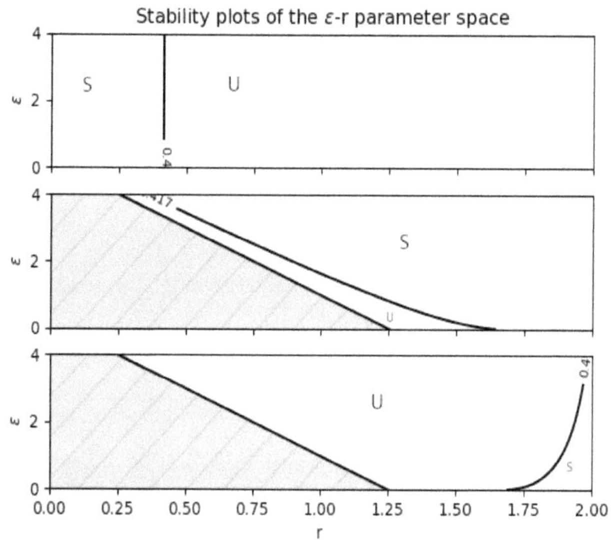

Figure 2.10: Existence and stability of equilibrium solutions in the (r,ε) parameter space. Top: trivial solution, middle: $(\tilde{X}_1, \tilde{Y}_1, \tilde{Z}_1)$, bottom: $(\tilde{X}_2, \tilde{Y}_2, \tilde{Z}_2)$. Linear stability regions denoted by S (stable) and U (unstable).

For the fixed parameters (ν, ρ, q) described above, the Routh-Hurwitz criterion in Eqs. (2.7)-(2.9) becomes

$$w < 7/2$$

$$w < 4/7$$

$$w < 5/4$$

$$w^2 - \frac{7}{2}w + \frac{9}{7} > 0$$

19

which reduces to

$$w < w_-$$

where

$$w_- = \frac{7}{4} - \frac{1}{4}\sqrt{\frac{199}{7}}.$$

Fig. 2.10 also shows the stability boundary for each of the trivial and nontrivial solutions

using $w = r - \varepsilon\left(2\tilde{Y} + 3\tilde{Y}^2\right)$. For the trivial solution, $\tilde{Y} = 0$ and so the stability boundary

is $r < w_-$ and is independent of the strength of the nonlinear terms ε. Using Eq. (2.4) for $\tilde{Y}$,

the stable region for the nontrivial solutions is as shown in Fig. 2.10 (middle) and (bottom).

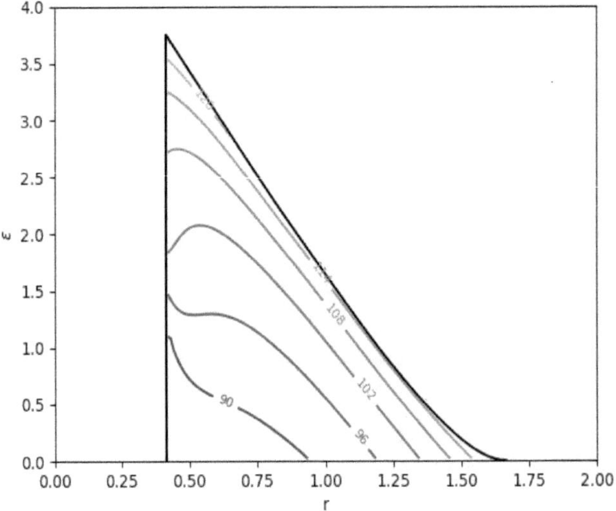

Figure 2.11: The region of limit cycles for the SM91 model in the (r,ε) parameter space. Contours are the dimensional period of limit cycles in kyrs using $t_c = 10$ kyrs [19].

The region in parameter space allowing limit cycles corresponds to the region where there

is no stable equilibria, as shown in Fig. 2.11 (like Figure 3 of [19]). The left boundary corresponds to the stability boundary of the trivial solution, the right boundary corresponds to the stability boundary of the (middle) nontrivial solution. We perform numerical calculations of the nondimensional system for (r, ε) throughout this region of Fig. 2.11 to determine if limit cycles exist, and if so, the period of the limit cycle. Existence of limit cycles is determined by monitoring the X component of the solution as a function of time, determining the period from the negative to positive zero crossings of the solution. The solution is tracked until the period converges to within 0.1%. Using this method, the period of limit cycle solutions can be determined on a grid covering the region in (r, ε) space, from which a contour plot of the period of the limit cycle solutions can be superimposed on Fig. 2.11. We plot the period in dimensional time using a timescale of $t_c = 10$ kyr as in [19]. Thus, limit cycles of ≈ 100 kyr do exist in the (r, ε) space, but they require the linear growth coefficient of the CO_2 (Y) equation r to be in a specific range, and the strength of the nonlinear terms ε to be linked to the value of r as shown by the 100 kyr contour.

Other ice age models

Other similar models that we will consider are the Saltzman and Sutera 1987 model (SS87) (reduced form, [20], p. 238):

$$\begin{aligned}
\frac{dI'}{dt} &= -a_0 I' - a_1 \mu' \\
\frac{d\mu'}{dt} &= b_1 \mu' + b_5 \theta' - b_6 \theta'^2 \mu' \\
\frac{d\theta'}{dt} &= c_0 I' - c_2 \theta',
\end{aligned} \tag{2.10}$$

a modification of the SS87 model (BR22):

$$\frac{dI'}{dt} = -a_0 I' - a_1 \mu'$$

$$\frac{d\mu'}{dt} = b_1 \mu' + b_5 \theta' - 2b_6 \theta' \mu' \qquad (2.11)$$

$$\frac{d\theta'}{dt} = c_0 I' - c_2 \theta',$$

and finally, a Saltzman and Maasch 1990 model (SM90) ([15], p. 1955):

$$\frac{dI'}{dt} = -a_0 I' - a_1 \mu' - a_2 M(t)$$

$$\frac{d\mu'}{dt} = b_1 \mu' - (b_2 - b_3 N')N' - b_4 N'^2 \mu' \qquad (2.12)$$

$$\frac{dN'}{dt} = c_0 I' - c_2 N',$$

where the variables and notations are similar to the ones in Eq. (2.1) except that in Eq. (2.12) the variable N denotes the "deep ocean warmth and salinity associated, for example, with North Atlantic Deep Water, or NADW," and M denotes external forcing like Milankovitch forcing, as stated in [15], p. 1955. The problem with these models, however, is that, again, they are not derived from governing equations from physics or chemistry. The theory of the Pleistocene Ice Ages is still in its early stages and as of 2013, there isn't a satisfying theory for it yet ([5], p. 300).

2.3 Introduction to a 2009 Crucifix paper

The main direction of the work in this dissertation is to develop and test an idea put

forth in a paper by M. Crucifix (2011) [3] and M. Crucifix and J. Rougier (2009) [4]. The

general idea of [4] (from now on be called "Crucifix & Rougier (2009)") is to use a particle

filter to estimate the parameters of a DE model in order to create a "best-fit" solution based

on measurements of a particular variable, using a stochastic forward Euler method called

the Euler-Maruyama method (EM). Thus, this combination of particle filter with EM will

give a Bayesian description of the parameters that adapts dynamically to the data as time

evolves. Long term changes in the parameters could be thought of as a response to tectonic

changes in the Earth as stated in [15]. The model that M. Crucifix and J. Rougier used was:

$$
\begin{aligned}
\frac{dI'}{dt} &= -a_1 \left[k_\mu \mu' + k_\theta \theta' + \kappa_R R'(t) \right] - K_I I' \\
\frac{d\mu'}{dt} &= b_1 \mu' - b_2 \mu'^2 - b_3 \mu'^3 - b_\theta \theta' \\
\frac{d\theta'}{dt} &= -c_1 I' - K_\theta \theta'
\end{aligned}
\tag{2.13}
$$

which is an equivalent form of the SM91 model, without stochastic forcing (Crucifix &

Rougier (2009), p. 15). The primes (') on the variables, again, represent the deviations from

some mean or fixed point (see [19], p. 203, equations (7) - (10)). In using a particle filter on

Eq. (2.13), the a_1, b_i, and c_1 will be probability distributions while the k_i and K_i will be fixed

based on geological estimates (Crucifix & Rougier (2009), Table 1, p. 24). Prior distributions

for the unknown parameters are assumed and there is an initial transient in which the prior

distribution is evolved by the response to the data. Crucifix & Rougier (2009) found that particle filter did track the CO_2 measurements reasonably well but as they said on p. 28, the "CO_2 dynamics during interglacial periods are not satisfactorily reproduced by this model," and it may be due to calibration issues with the parameters. In particular, on p. 25 - 27 of Crucifix & Rougier (2009),

> ... model predictions for climate evolution after -126 kyr are overall consistent with data. These predictions are shown by the thin gray lines. They are obtained as model realisations sampled from the latest state and parameter estimates at -126 kyr. In particular the model succeeds in predicting the draw-down in CO_2 and the large ice mass increase around -70 kyr. Model predictions then split but 60 % are consistent with ice volume above $20 \ 10^{18}$ kg of ice at the last glacial maximum, and 36 % are consistent with the more stringent prediction of ice volume above $30 \ 10^{18}$ kg of ice at that time. By contrast, the late decrease in CO_2 around -90 kyr as compared to the data constitutes a severe weakness. The relatively weak predictability of CO_2 is also noteworthy.

The Crucifix & Rougier (2009) paper will motivate the work in this dissertation in that it will give us a baseline to compare our results to. The issue that we would like to address is that we would like to verify that the "particle filter" fails to discriminate models based on their long term dynamics as claimed in [3], p. 839.

2.4   Introduction to Particle Filtering Methods

Particle filter methods are "sequential Monte Carlo methods based on point mass (or "particle") representations of probability densities, which can be applied to any state-space model and which generalize the traditional Kalman filtering methods" ([1], p. 174). There are many different types of algorithms depending on the problem (see [1], p. 177 - 183), but in this dissertation we will focus mainly on the Liu and West particle filter (LWPF) [12], which allows for filtering on state variables and parameters of the system. The LWPF framework is as follows (following similar notation as in [12]): a sample of state variables x_t and parameters θ_t at time t,

$$\{x_t^j, \theta_t^j : j = 1, ..., M\}$$

and their corresponding weights

$$\{\omega_t^j : j = 1, ..., M\}$$

are used as an importance sample approximation to the posterior distribution $p(x_t, \theta | D_t)$, where D_t are the measurements y_1, y_2, ..., y_t. It should also be noted that the measurement error is normal with mean 0 and standard deviation sdy, for some sdy > 0. The algorithm for LWPF is outlined as follows (see [12], p. 207):

1. From the prior point estimates of (x_t, θ_t) compute

$$\mu_{t+1}^j = E\left(x_{t+1} | x_t^j, \theta_t^j\right) \tag{2.14}$$

$$m_t^j = a\theta_t^j + (1-a)\bar{\theta}_t \tag{2.15}$$

$$\bar{\theta}_t = \frac{1}{M}\sum_{j=1}^{M}\theta_t^j$$

for $j = 1, ..., M$, where the expected value E is a weighted average with weights ω_t^j and $a > 0$

is related to the kernel parameter that will be defined later.

2. Sample an auxillary integer variable $k \in \{1, ..., M\}$ with probabilities proportional to

$$g_{t+1}^{j} \propto \omega_{t}^{j} p\left(y_{t+1} | \mu_{t+1}^{j}, m_{t}^{j}\right) \tag{2.16}$$

3. Sample a new parameter vector θ_{t+1}^{k} such that

$$\theta_{t+1}^{k} \sim N\left(m_{t}^{k}, h^{2} V_{t}\right) \tag{2.17}$$

where V_t is the Monte Carlo variance of a given probability density function $p\left(\theta | D_t\right)$ and h is the kernel smoothing parameter such that

$$h = \sqrt{1 - a^2} \tag{2.18}$$

$$a = \frac{3\delta - 1}{2\delta} \tag{2.19}$$

where $\delta \in (0, 1]$ is a discount factor that is typically between $0.95 - 0.99$ (see [12], section 10.3.3, p. 204 - 206).

4. Sample a new value of the current state vector x_{t+1}^{k} from the process model as

$$p\left(x_{t+1}^{k} | x_{t}^{k}, \theta_{t+1}^{k}\right) \tag{2.20}$$

5. Evaluate the corresponding weight

$$\omega_{t+1}^{k} \propto \frac{p\left(y_{t+1} | x_{t+1}^{k}, \theta_{t+1}^{k}\right)}{p\left(y_{t+1} | \mu_{t+1}^{k}, m_{t}^{k}\right)} \tag{2.21}$$

6. Repeat steps (2) - (5) until one produces a final posterior approximation $\left(x_{t+1}^k, \theta_{t+1}^k\right)$ with weights ω_{t+1}^k as needed.

Liu and West's choice of h is motivated by reducing the variance in the probability distribution for the parameters.

There are some R packages that utilize the LWPF in some form like the *bsmc2* [9] (a modified version of the LWPF), but there doesn't appear to be any public-domain versions of the LWPF written in Python.

In this paper, we will create and implement a version of the LWPF on different Saltzman-type DE models. Our master code is given in Appendix A. We provide more detail about the numerical implementation and code validation in Chapter 3. In Appendix H we provide a "proof of concept" application of the particle filter to demonstrate how it can determine the parameters for the Lorenz system from measuring one component of the solution.

Chapter 3

Numerical Methods and Results

3.1   Numerical method and validation

Before we present any benchmarking results of the particle filter that will be used, we will first validate our numerical ODE solver using the nonstochastic and stochastic forward Euler methods (see Appendix A, line 307, "fwdEuler_rand2D" function). We compare results directly to Crucifix & Rougier (2009) using the numerical results in their Fig. 7. The comparison uses a nonstochastic and stochastic forward Euler on a Saltzman-Maach 1991 model [19] (SM91) with and without insolation forcing; the timestep used was 100 years and the parameter values used are listed in Table 3.1; the algorithm was initialized with $I_0 = 4.4$, $\mu_0 = -9.0$ and $\theta_0 = -0.22$ and then ran from -450 to 0 kiloyears.

Table 3.1: Parameter values used in the SM91 model. See Crucifix & Rougier (2009), Table 1, p. 24.

Fixed (non-stochastic) model parameters:

k_μ	$11/253$	$^\circ$C/ppmv
k_R	0.08	$10^{18}{}^\circ$C/(Wm^{-2})
k_θ	0.5	$^\circ$C/$^\circ$C
K_I	10^{-4}	$1/$yr
K_θ	2.5×10^{-4}	$1/$yr

Stochastic model parameters:

a_1	8.7×10^{-4}	10^{18} kg/$^\circ$C yr
b_1	1.3×10^{-4}	$1/$yr
b_2	1.1×10^{-6}	$1/$(ppm yr)
b_3	3.6×10^{-8}	$1/$(ppm^2 yr)
b_θ	5.6×10^{-3}	ppm/($^\circ$C yr)
c_1	1.2×10^{-5}	$^\circ$C/(10^{18}kg yr)

where each sampled parameter in the particle filter will have a log-normal distribution with mean equal these values and a standard deviation of 20 % of the mean.

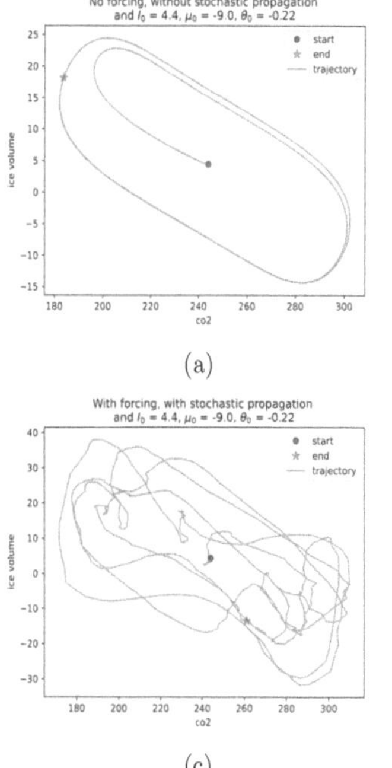

(a)

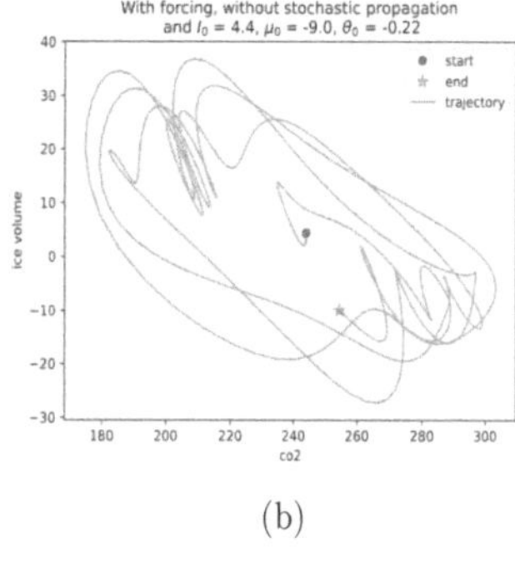

(b)

(c)

Figure 3.1: Plots of phase portraits of the ice volume and CO$_2$ for the SM91 model (a) without external forcing, (b) with external forcing, (c) with external forcing and noise.

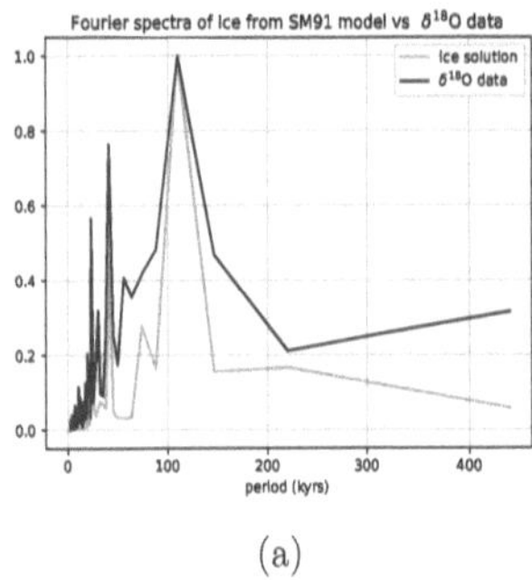
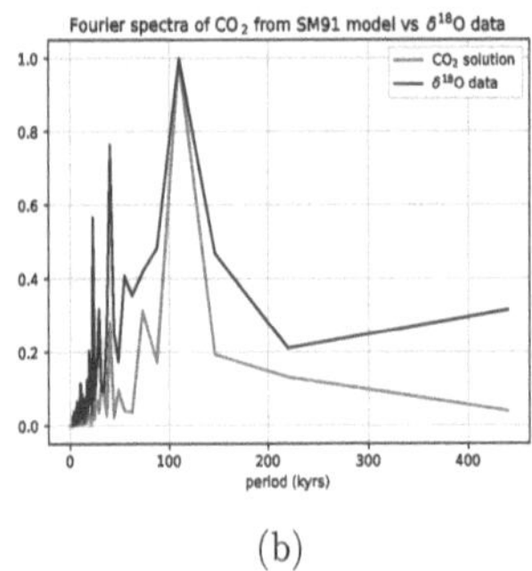

(a)   (b)

Figure 3.2: Plots of the Fourier power spectra vs period of the past 440 kyrs of the: (a) ice synthetic data and δ^{18}O data [24], (b) CO_2 synthetic data and δ^{18}O data [24].
Note: Fourier spectra are normalized with respect to the peak power.

Fig. 3.1 shows our numerical results as a projection of the (I, μ, θ) solution trajectory onto the 2D (I, μ) space. As shown in Fig. 3.1, (a) and (b) are nearly identical to the first two figures in Figure 7 in Crucifix & Rougier (2009). For Fig. 3.1 (c), stochastic propagation with variance

$$\Sigma = \frac{1}{10^2} \begin{pmatrix} 0.5^2 & 0 & 0 \\ 0 & 5.0^2 & 0 \\ 0 & 0 & 0.5^2 \end{pmatrix}$$

was used. The variance matrix Σ enters into the differential equation as (using similar notation as in Crucifix & Rougier, p. 19):

$$d\mathbf{x} = \mathbf{f}(\mathbf{x})\, dt + \sqrt{\Sigma}\, d\mathbf{W} \tag{3.1}$$

where $\mathbf{x}$ is the vector of ice, CO_2 and temperature, $\mathbf{f}(\mathbf{x})$ is the "right-hand side" of Eq. (2.13), and $\mathbf{W}$ is the standardized Brownian motion (i.e. Brownian motion with mean 0 and variance

30

1). Our results in Fig. 3.1(c) produced a figure that is not exactly the same as the third figure in Figure 7 in Crucifix & Rougier (2009) but is qualitatively similar. This should be expected due to the very nature of the Euler-Maruyama method as it depends on random stochastic forcing (i.e. no two simulations will be the same). Therefore, overall, one can be confident in the output of our numerical ODE solver for the (non-stochastic) forward Euler or (stochastic) Euler-Maruyama cases.

Returning to our discussion of the ice age data having a strong 100 kyr signal that is not present in the astronomical forcing, Fig. 3.2 is a graph of the Fourier spectrum of the synthetic data for ice (Fig. 3.2(a)) and CO_2 (Fig. 3.2(b)). The graphs show that the SM91 model with the chosen parameters can give a strong 100 kyr signal as observed in the $\delta^{18}O$ data, and thus may provide a solution to the 100 kyr problem.

Now, we move on to a series of benchmarks for determining how well the particle filter does on synthetic data. The synthetic data was generated using a nonstochastic forward Euler method on the SM91 model with insolation forcing, using the same time step, parameter values and time interval as before except that the initial conditions are now $I_0 = 4.0$, $\mu_0 = -9.0$ and $\theta_0 = -0.22$. In the particle filter, we will use the synthetic CO_2 data as input and the "observation error" (sdy) as 5 ppmv instead of 20 ppmv (see Crucifix & Rougier (2009), p. 25) because, as we will show in the following results, we found that this value yielded a better error profile. In addition to using the Liu and West particle filter [12] we incorporate a "resampling threshold" (rs_thresh) to help prevent degeneracy problems (see [1], p. 180, algorithm 3; in the algorithm we use (rs_thresh) (number of samples) in place of N_T). The resampling threshold is implemented in our algorithm as follows: given a resampling threshold $0 < \text{rs\_thresh} \leq 1$ and the auxiliary particle weights $g_{t+1}^{(j)}$ as calculated from

Eq. (2.16),

---

1. Calculate: $g_{\text{sum}}^{t+1} = \sum_{j=1}^{M} g_{t+1}^{(j)}$.

2. Normalize each $g_{t+1}^{(j)}$: $\tilde{g}_{t+1}^{(j)} = g_{t+1}^{(j)} / g_{\text{sum}}^{t+1}$.

3. Calculate N_{eff}: $N_{\text{eff}} = 1/\sum_{j=1}^{M} \left(\tilde{g}_{t+1}^{(j)} \right)^2$.

4. If $N_{\text{eff}} <$ (nsamp) (rs_thresh), then resample the index k from step 2 of Eq. (2.16).

Otherwise, k will remain unchanged.

Instead of fixing the covariance matrix Σ as was done in Crucifix & Rougier (2009), here we will consider different cases to see how well the particle filter does in each case. Throughout this paper we will define two choices of Σ as sdx1 or sdx2 where

$$ \text{sdx1} = \frac{1}{100^2} \begin{pmatrix} 0.5^2 & 0 & 0 \\ 0 & 5.0^2 & 0 \\ 0 & 0 & 0.5^2 \end{pmatrix} $$

and

$$ \text{sdx2} = \frac{1}{10^2} \begin{pmatrix} 0.5^2 & 0 & 0 \\ 0 & 5.0^2 & 0 \\ 0 & 0 & 0.5^2 \end{pmatrix} $$

as the two different covariance matrices that will be used to produce different results for comparison.

In order to see how well the algorithm does, we defined a measure of the error E as an average of a weighted L^1 error between the measured CO_2 values and the calculated ones from the particle filter, x, over time $t = t_1, t_2, ..., t_N$. More specifically,

$$E = \frac{1}{N} \sum_{m=1}^{N} err(m) \tag{3.2}$$

where

$$err(m) = \sum_{k=1}^{n_{\text{samp}}} \frac{|y_{\text{true}}^m - x_k^m|}{|y_{\text{true}}|_{\text{avg}}} w_k \tag{3.3}$$

with

$$|y_{\text{true}}|_{\text{avg}} = \frac{1}{N} \sum_{m=1}^{N} |y_{\text{true}}^m| \tag{3.4}$$

and

$$N = \text{total number of time steps}$$

$$n_{\text{samp}} = \text{number of samples}$$

$$y_{\text{true}}^m = \text{true or measured value of y at time } t_m$$

$$x_k^m = k^{th} \text{ sample of the approximate y value at the } m^{th} \text{ time step}$$

$$w_k = k^{th} \text{ weight associated with the } k^{th} \text{ sample}$$

A given particle filter simulation generates the relative L^1 error vs time, err(m), and the total integrated error E. To determine how the accuracy depended on the algorithm parameters, we conducted multiple runs using:

$$\text{rs\_thresh} = 0.1 - 1.0, \text{ with a step size of } \Delta(\text{rs\_thresh}) = 0.1$$

$$\text{sdy} = 5 - 20, \text{ with a step size of } 5$$

$$\text{nsamp} = 10 - 100, \text{ with a step size of } \Delta(\text{nsamp}) = 5, \text{ then nsamp} = 500$$

where each run is repeated over some number of trials (Ntrials). Here Ntrials will be 30 so that the plots of error bars of E will be a "good enough" representation of the error for a given sample. Our benchmarking results are detailed in Appendix B. Using the metric as defined in Eq. (3.2), we found that sdy $= 5.0$ and rs_thresh $= 1.0$ yielded the best E profile in comparison to the other choices of sdy and rs_thresh (in terms of how "low" the curve of E was) and therefore we will fix these values for the remainder of the report.

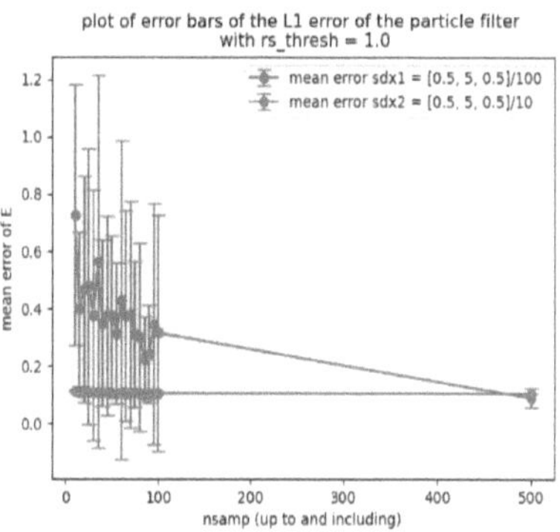
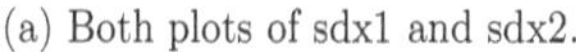

(a) Both plots of sdx1 and sdx2.

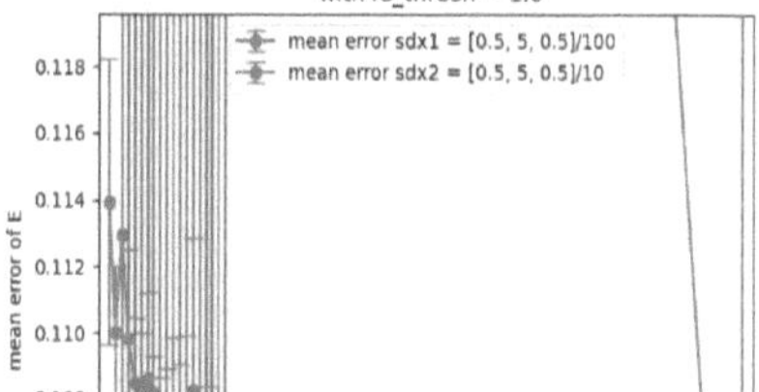

(b) Close-up view of the plot corresponding to sdx2.

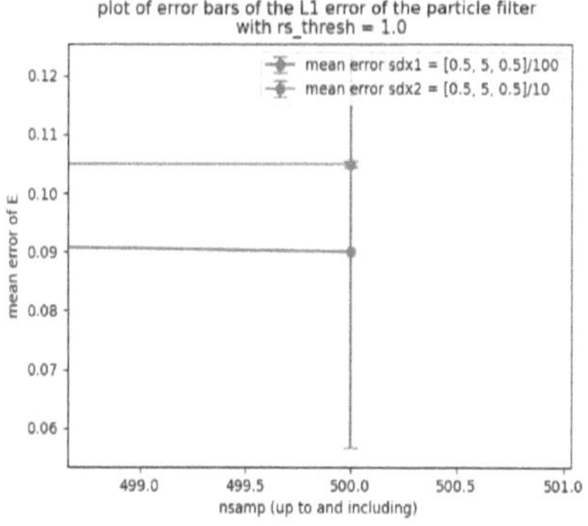

(c) A zoomed-in picture of both plots corresponding to nsamp = 500.

Figure 3.3: Plots of the error bars of E using two different process errors sdx1 and sdx2. Error bars are based on standard deviation of the results from Ntrial = 30 trials for each point.

Fig. 3.3 (a) is a plot of the overall error E as a function of nsamp for two different values of Σ. We see that the overall trend for both plots of E is decreasing, implying that as the number of samples increases the error decreases. Interestingly though, the green plot of E for sdx2 already appears to be decreasing more slowly around nsamp = 100 as shown in Fig. 3.3 (b), whereas the red plot for sdx1 still appears to have a negative slope at nsamp =

500. As shown in Fig. 3.3 (c), we can see that the difference in E between the two at nsamp
= 500 is relatively small but the sdx1 results (red) has much more variation than the sdx2
results (green).

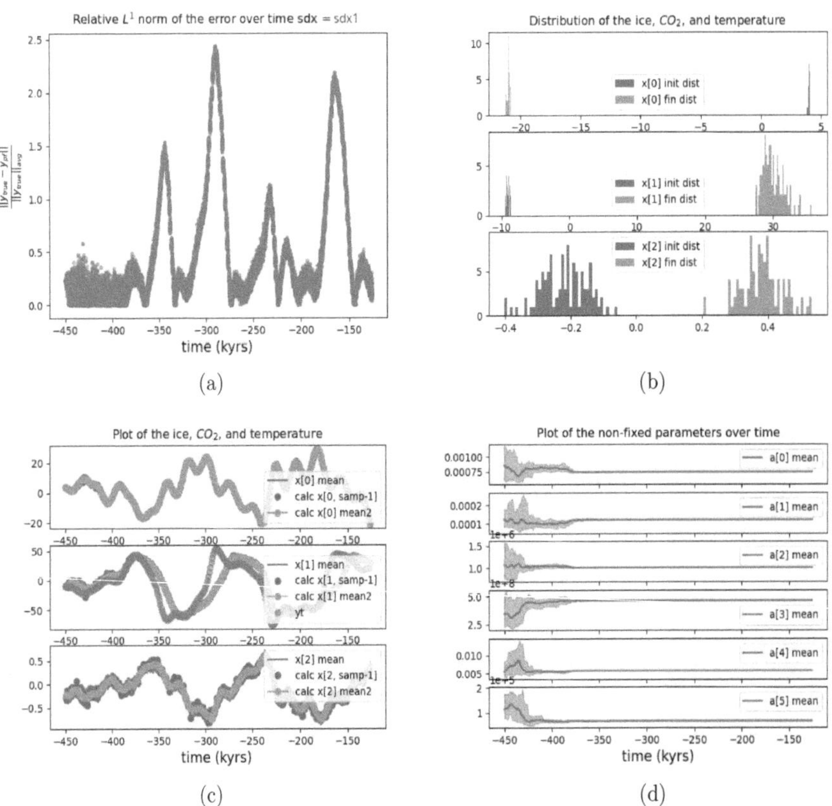

Figure 3.4: Particle filter results for Σ = sdx1 and nsamp = 100: (a) relative L^1 error vs
time, (b) distribution of state variables at initial and final time, (c) evolution of state variable
distributions vs time, (d) evolution of parameter distributions vs time.

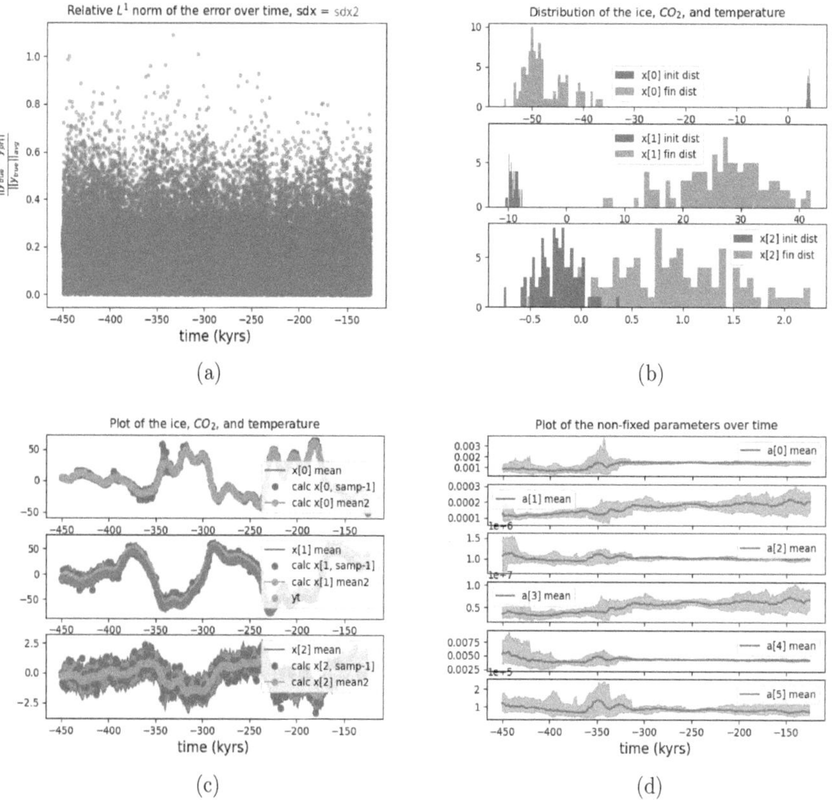

Figure 3.5: Particle filter results for $\Sigma = \text{sdx2}$ and $\text{nsamp} = 100$: (a) relative L^1 error vs time, (b) distribution of state variables at initial and final time, (c) evolution of state variable distributions vs time, (d) evolution of parameter distributions vs time.

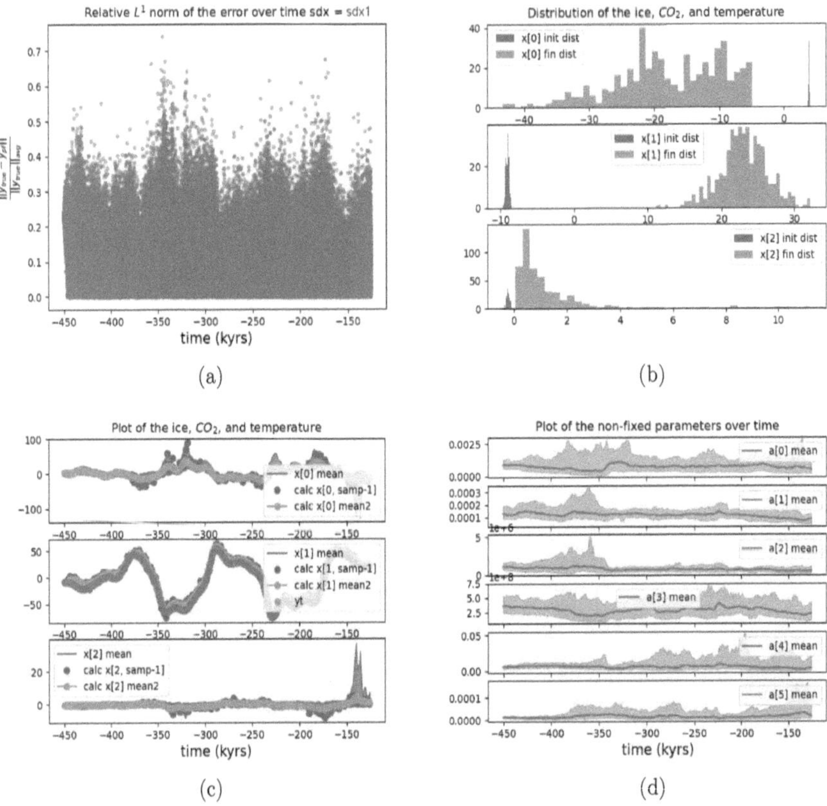

Figure 3.6: Particle filter results for $\Sigma = \mathrm{sdx1}$ and $\mathrm{nsamp} = 500$: (a) relative L^1 error vs time, (b) distribution of state variables at initial and final time, (c) evolution of state variable distributions vs time, (d) evolution of parameter distributions vs time.

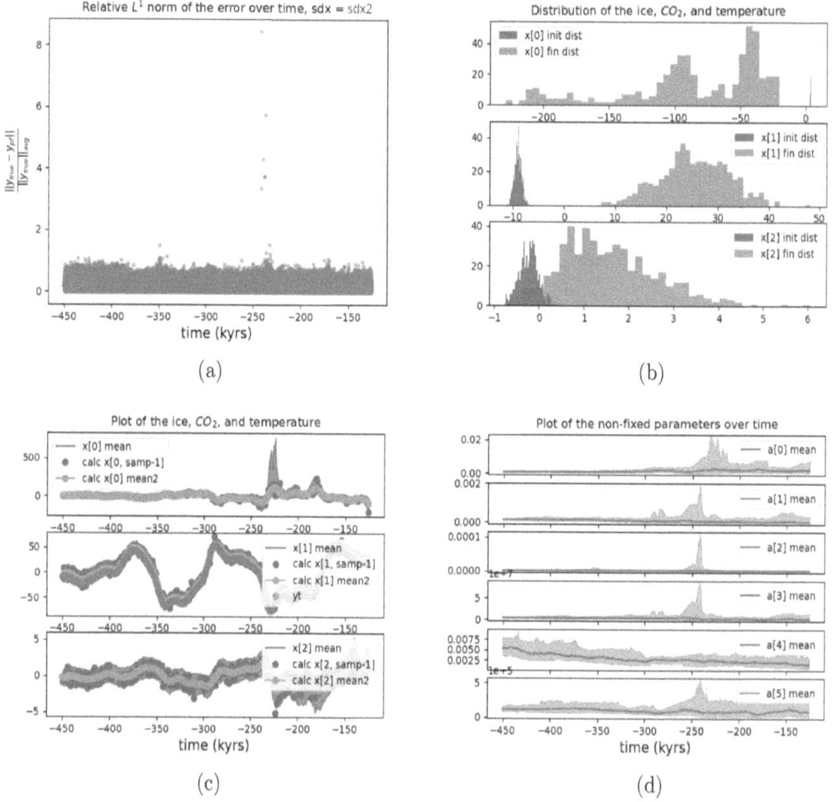

Figure 3.7: Particle filter results for $\Sigma = $ sdx2 and nsamp $= 500$: (a) relative L^1 error vs time, (b) distribution of state variables at initial and final time, (c) evolution of state variable distributions vs time, (d) evolution of parameter distributions vs time.

Next, to get a better idea of how well the particle filter is doing for each Σ, we looked at

the distributions of the state and parameter values and their plots over time while varying the

number of samples. To understand what Fig. 3.4 (and similarly others) are showing, (a) is a

plot of the relative error between the exact and the particle filter results of the CO_2 values;

(b) is showing the initial and final distributions of the ice, CO_2, and ocean temperature

anomalies; (c) shows the particle filter results of the last sample (blue dots) with the average

39

value of the sample (orange) and a 95% confidence interval (light blue shade) along with the observed data (green); (d) is a plot of the parameters over time with the average value of the particle filter sample (red) and its 95% confidence interval (light green shade). Fig. 3.4 (sdx1) and Fig. 3.5 (sdx2) give the plots over time as well as their distributions for nsamp = 100, whereas Fig. 3.6 (sdx1) and Fig. 3.7 (sdx2) are the same thing except nsamp = 500. Notice that there is more variability in the confidence interval in Fig. 3.5 (d) versus Fig. 3.4 (d) and that the CO_2 data lies more within the confidence interval in Fig. 3.5 (c) than in Fig. 3.4 (c), which may explain why the green plot (sdx2) in Fig. 3.3 (a) does better than the red one (sdx1). However, as shown in Fig. 3.3 (c), the red and green curves are not much different from one another and comparing Fig. 3.6 (c) and Fig. 3.7 (c) have similar CO_2 plots as well.

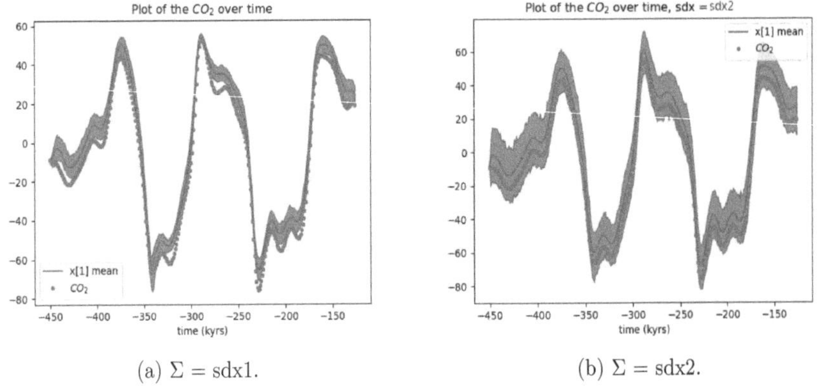

(a) $\Sigma = \text{sdx1}$.

(b) $\Sigma = \text{sdx2}$.

Figure 3.8: Plots of the CO_2 measurements and the mean of the particle filter samples of the CO_2, for nsamp = 500.

Taking a closer look at the CO_2 values in Fig. 3.8, we can see that the CO_2 data (green) lies more in the confidence interval in Fig. 3.8 (b) than in Fig. 3.8 (a), and that the mean of the particle filter data (red) appears to follow the data more in Fig. 3.8 (b) than in Fig. 3.8

(a).

To further confirm the trend that the red curve in Fig. 3.3 (a) continues to decrease more than the green curve, error data for nsamp = 200 – 1000 in steps of 100 was produced (again using 30 trials to determine the error bars) and is shown in Fig. 3.9.

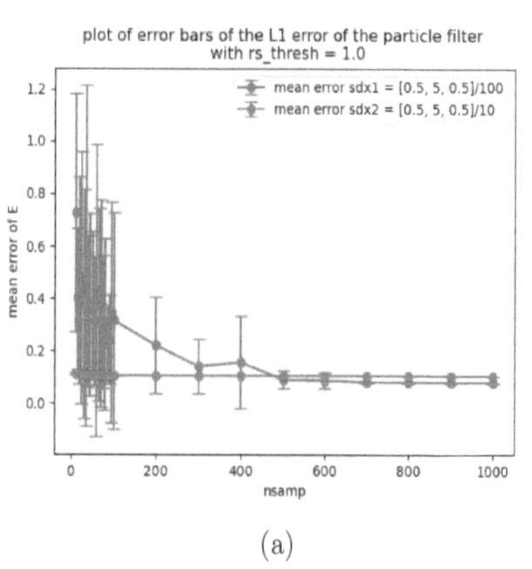

(a)

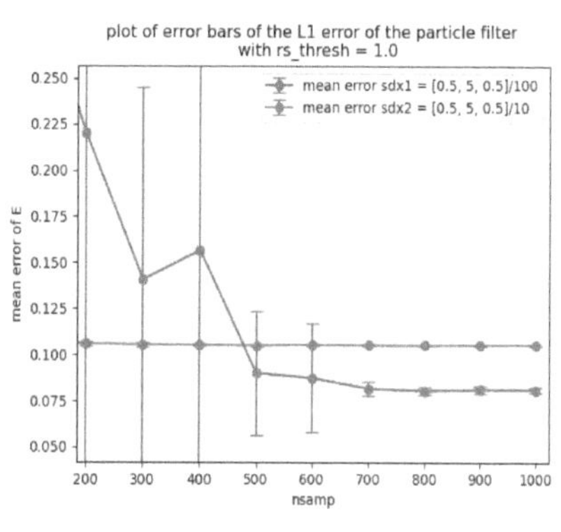

(b) A close-up view of (a) over nsamp = 200 - 1000.

Figure 3.9: Graph of overall error E vs nsamp (as in Fig. 3.3 except that it includes more points for nsamp = 200 - 1000).

In Fig. 3.9 (a) one can see that the green curve doesn't decrease as much as the red curve (although the red curve appears to be approaching an asyptote). In both curves there appears to be a "diminishing return" in terms of the measure E relative to nsamp, where both curves appear to have an asymptote.

We model the asymptotic behavior using an exponential model. The models were generated using the nonlinear least squares method "Levenberg-Marquardt algorithm as implemented in MINPACK", as stated in [21]. The following code was used to generate the models:

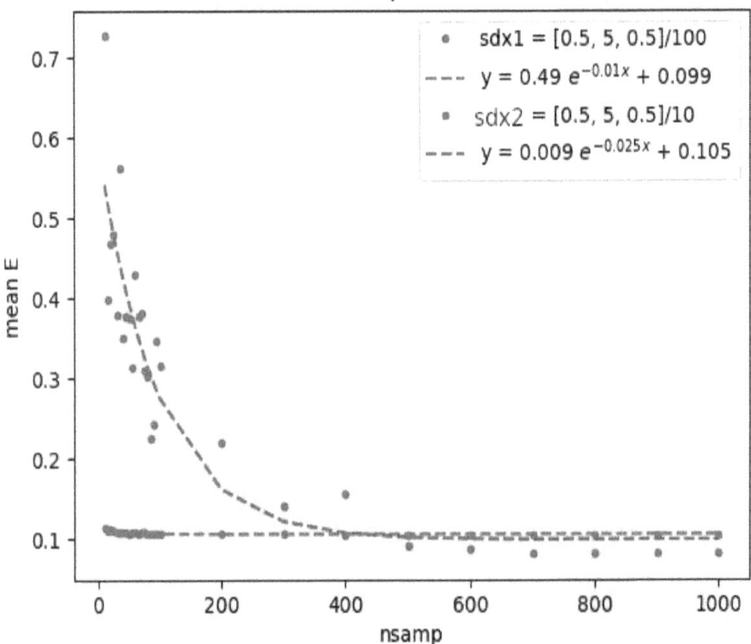

Figure 3.10: The best fit exponential models of the mean of E.

```python
import scipy.optimize as so
import numpy as np

# The proposed cost function (my guess):
# (see https://scipy-cookbook.readthedocs.io/items/robust_regression.html)
def func(coeffs, *args):
  nsamp, error = args
  return coeffs[0]*np.exp(coeffs[1]*nsamp) + coeffs[2] - error

def model(coeffs, nsamp):
  return coeffs[0]*np.exp(coeffs[1]*nsamp) + coeffs[2]


# initial guess:
print('mean1\n')
coeffs0 = np.array([1.2, -0.05, 1.5], dtype=np.float64)
res_lsq1 = so.least_squares(func, coeffs0, args=(nsamp_arr, E_arr_mean11),
    method='lm', ftol=1e-15, loss='linear')
print("Success:",res_lsq1.success)
print(res_lsq1.x)
print()
```

```python
21
22 # initial guess:
23 print('mean2\n')
24 coeffs0 = np.array([0.5, -0.5, 1.5], dtype=np.float64)
25 res_lsq2 = so.least_squares(func, coeffs0, args=(nsamp_arr, E_arr_mean21),
        method='lm', ftol=1e-15, loss='linear')
26 print("Success:",res_lsq2.success)
27 print(res_lsq2.x)
28 print()
```

where "nsamp_arr" is the array of nsamp values and "E_arr_mean" is array of the average E values for the different nsamp values. The results of the exponential fit for each model are shown in Fig. 3.10. Both models show that the limiting error is achieved on the order of nsamp $\approx$ 100 - 500. Armed with this knowledge, one can argue that it isn't necessary to have, say, nsamp = 50,000 if the difference in the average E values between nsamp = 1,000 and 50,000 is only, say, 1%, offering a chance for a significant speed up. One can show that in order to be within 1% of the asymptotes of the models depicted in Fig. 3.10, $x \approx 620.4$ for sdx1 and $x \approx 85.9$ for sdx2. Thus, using nsamp = 621 for sdx1 and nsamp = 86 for sdx2, Fig. 3.11 and Fig. 3.12 were produced.

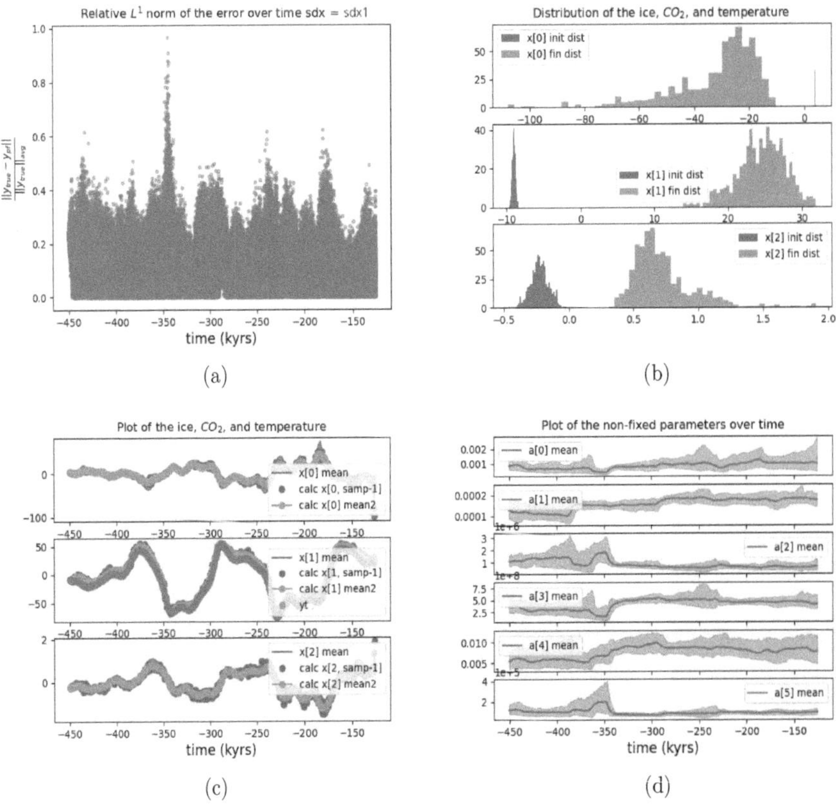

Figure 3.11: Particle filter results for $\Sigma = $ sdx1 and nsamp $= 621$: (a) relative L^1 error vs time, (b) distribution of state variables at initial and final time, (c) evolution of state variable distributions vs time, (d) evolution of parameter distributions vs time.

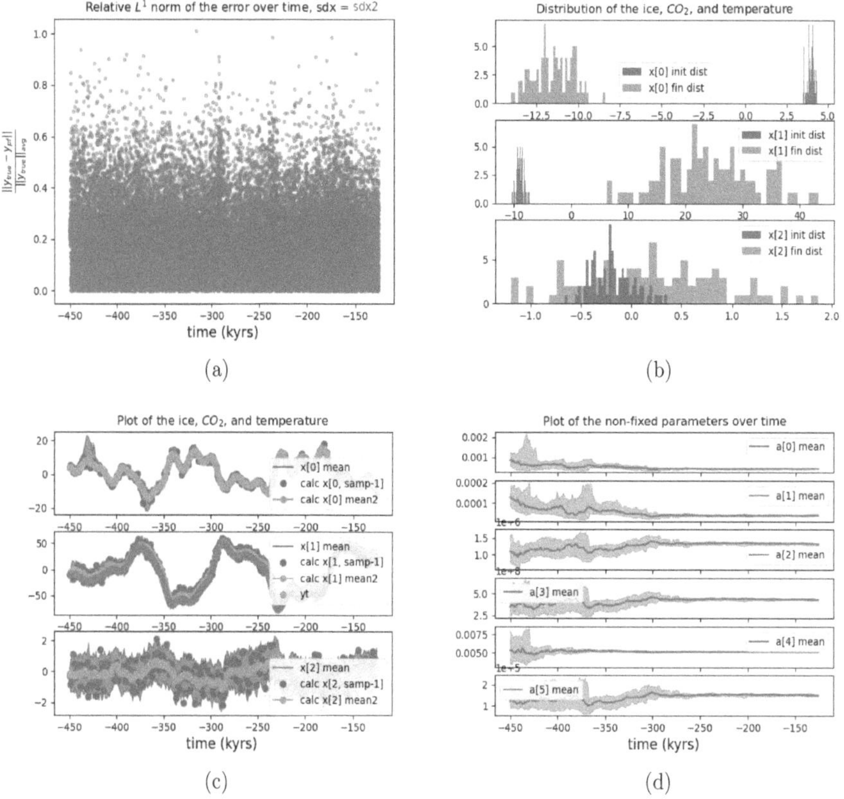

Figure 3.12: Particle filter results for $\Sigma =$ sdx2 and nsamp $= 86$: (a) relative L^1 error vs time, (b) distribution of state variables at initial and final time, (c) distribution of parameters of initial and final time, (d) evolution of state variable distributions vs time, (e) evolution of parameter distributions vs time.

The statistics of the final parameter values for $\Sigma =$ sdx1 and $\Sigma =$ sdx2 are listed in Table 3.2. The exact values in Table 3.2 are the same values in Table 3.1 since the particle filter is on the synthetic CO_2 data and (ideally) the mean of the sample of parameters from the particle filter shouldn't be too far off from the initial parameter values from Table 3.1. Some of the average values of the parameters in Table 3.1 for sdx1 seem to be somewhat better than for sdx2, and similarly some of the average values of the parameters for sdx2

45

Table 3.2: Average parameter values found by the particle filter using the SM91 model on the synthetic data using $\Sigma = \text{sdx1}$ with nsamp = 621, and $\Sigma = \text{sdx2}$ with nsamp = 86.

param	exact	mean (sdx1)	rel err (sdx1)	mean (sdx2)	rel err (sdx2)
a_1	8.7×10^{-4}	9.581×10^{-4}	1.0×10^{-1}	4.440×10^{-4}	4.9×10^{-1}
b_1	1.3×10^{-4}	1.747×10^{-4}	3.4×10^{-1}	4.331×10^{-5}	6.7×10^{-1}
b_2	1.1×10^{-6}	6.232×10^{-7}	4.3×10^{-1}	1.335×10^{-6}	2.1×10^{-1}
b_3	3.6×10^{-8}	4.026×10^{-8}	1.2×10^{-1}	4.324×10^{-8}	2.0×10^{-1}
b_θ	5.6×10^{-3}	7.768×10^{-3}	3.9×10^{-1}	5.183×10^{-3}	7.4×10^{-2}
c_1	1.2×10^{-5}	9.363×10^{-6}	2.2×10^{-1}	1.485×10^{-5}	2.4×10^{-1}

seem to do better than sdx1, based on their relative errors. There doesn't seem to be a clear relationship why this is so and is most likely due to the stochastic nature of the DE system.

The benchmark results from this section are exclusively for the case of model selection using synthetic CO_2 data generated by the model. We now turn to using actual CO_2 data as a test of the model.

3.2   Parameter fitting on real data

The real data we use for our model fitting with the particle filter method is the CO_2 values from the Vostok, Antarctica data [14], which ranges from -440 to 0 kyrs.

The particle filter setup is similar to that of the previous section. The code uses a combination of a stochastic forward Euler on a Saltzman-Maach 1991 model with insolation forcing (June insolation at 65° N using a program provided in [10]), using a time step of 100 years in between measurements, but a critical difference is that it performs the particle filter algorithm only at each measured CO_2 value. These CO_2 measurements are not uniformly spaced in time. To accommodate the nonuniform CO_2 measurements the uniform dt time stepping was modified so that a smaller dt was used in the last step to give a time matching that of the measurement. The parameter values were initialized using the central parameter values as listed in Table 3.1 as in Crucifix & Rougier (2009), table 1 – item 1; the initial conditions used here were $I_0 = 15.0$, $\mu_0 \approx 51$ (initial CO_2 measured value minus it's mean) and $\theta_0 = 0.0$ and the algorithm ran from -450 to 0 kiloyears. As was done in Section 3.1, the particle filter used the same values as listed in our Table 3.1 (from Crucifix & Rougier (2009)), except based on our benchmarking of Section 3.1 we considered the "measurement noise" (sdy) as 5 ppmv instead of 20 ppmv (see Crucifix & Rougier (2009), p. 25) and we used a "resampling threshold" (rs_thresh) within the algorithm as discussed in Section 3.1. In addition, the values of the covariance matrix for the process model as in Table 1 of Crucifix & Rougier (2009) will also vary using either $\Sigma = $ sdx1 or $\Sigma = $ sdx2 to see how the particle filter will perform in each case.

We conducted an error analysis similar to that in Section 3.1 and we found that sdy $=$ 5.0 and an rs_thresh $=$ 1.0 gave the best E profile (just as we found in the benchmarking Section 3.1). Full details of this error analysis are given in Appendix C.

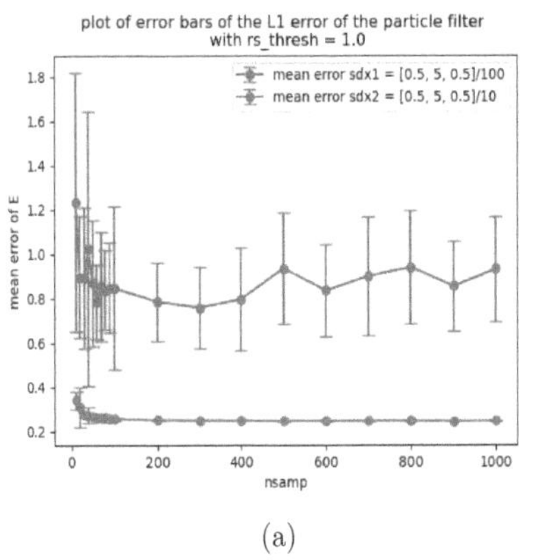
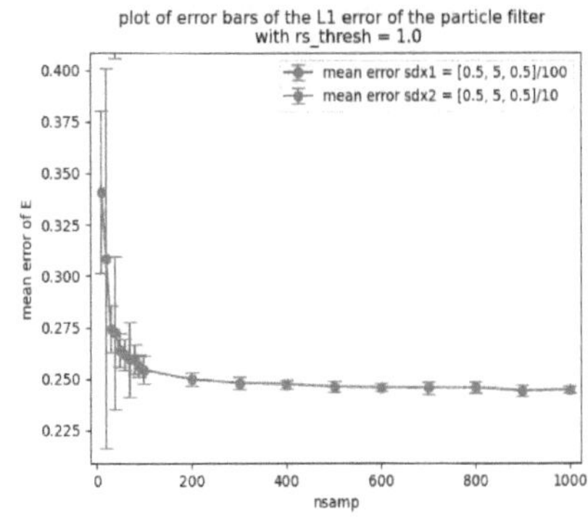

(a)

(b) Close-up view of the plot corresponding to sdx2 in (a).

Figure 3.13: Plots of the error bars of E using two different process errors sdx1 and sdx2.

Fig.3.13 presents the error results using sdy = 5 and rs_thresh = 1.0 for the two choices of Σ = sdx1 and Σ = sdx2. Looking at Fig. 3.13 (a), unlike what happened in Section 3.1, we see that the error corresponding to sdx2 is uniformly lower than the error data corresponding to sdx1. As shown in Fig. 3.13 (b), the trend for sdx2 appears to be decreasing exponentially so an exponential curve was calculated (using the same algorithms as before) to fit the data (see Fig. 3.14).

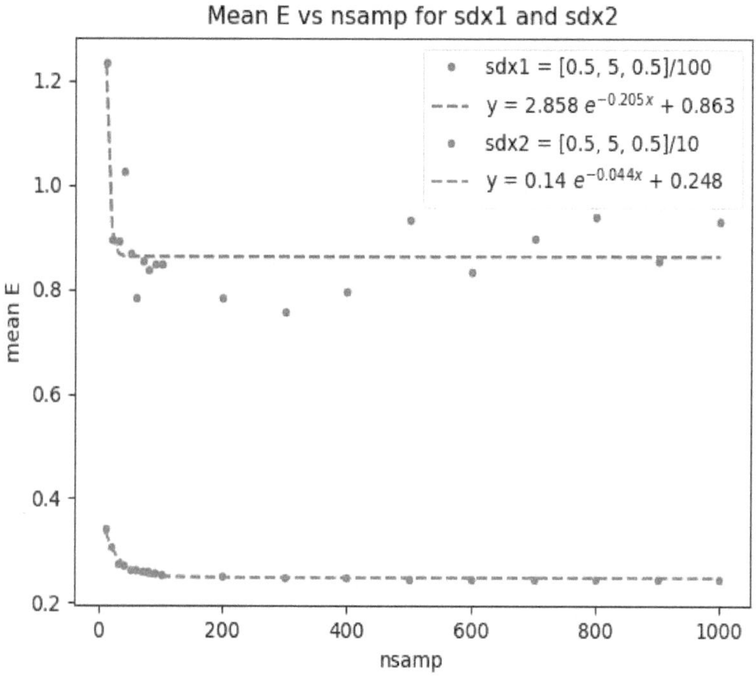

Figure 3.14: Exponential models for the mean of E for sdx1 and sdx2.

Using the exponential model for sdx2 from Fig. 3.14, and if we consider when E is 1% above its asymptote, then nsamp is about 92. However, for sdx1, the errors are significantly larger and have large error bars.

If we let nsamp = 10,000, Ntrials = 3, sdy = 5.0 and rs_thresh = 1.0, there still is no improvement in E for sdx1 as depicted in Fig. 3.15. Granted the number of trials was only 3 (this was chosen due to the limitations of memory and time), so the amount of variation is large.

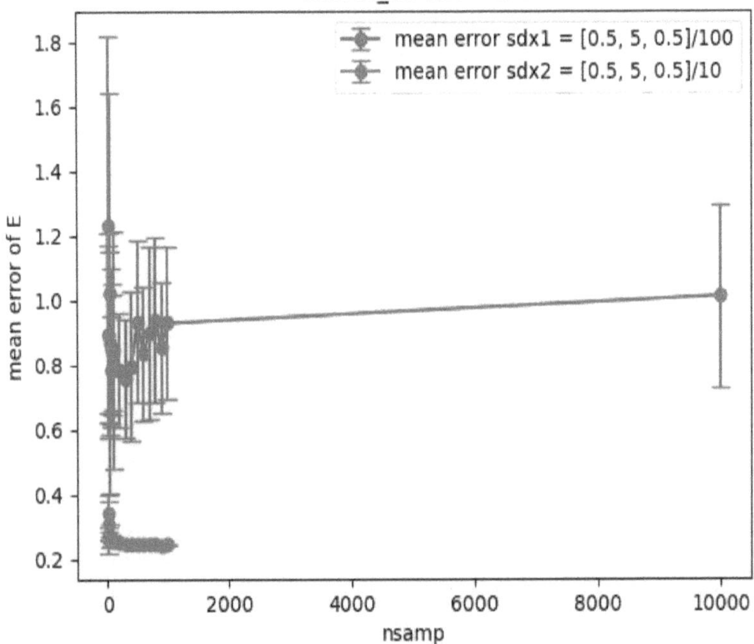

Figure 3.15: A continuation of Fig. 3.13.

Note: there is no E data for nsamp = 10000 for sdx2 because the algorithm developed numerical instability.

To further verify that there won't be much improvement in E for sdx1 in comparison to sdx2, we'll let nsamp = 50000 (as done in Crucifix & Rougier (2009)) for sdx1 and nsamp = 92 (as calculated above) for sdx2, and create the five plots as was done in Section 3.1 looking at various aspects of the model and parameter data. Fig. 3.16 shows the results for sdx1 and Fig. 3.17 show the results for sdx2.

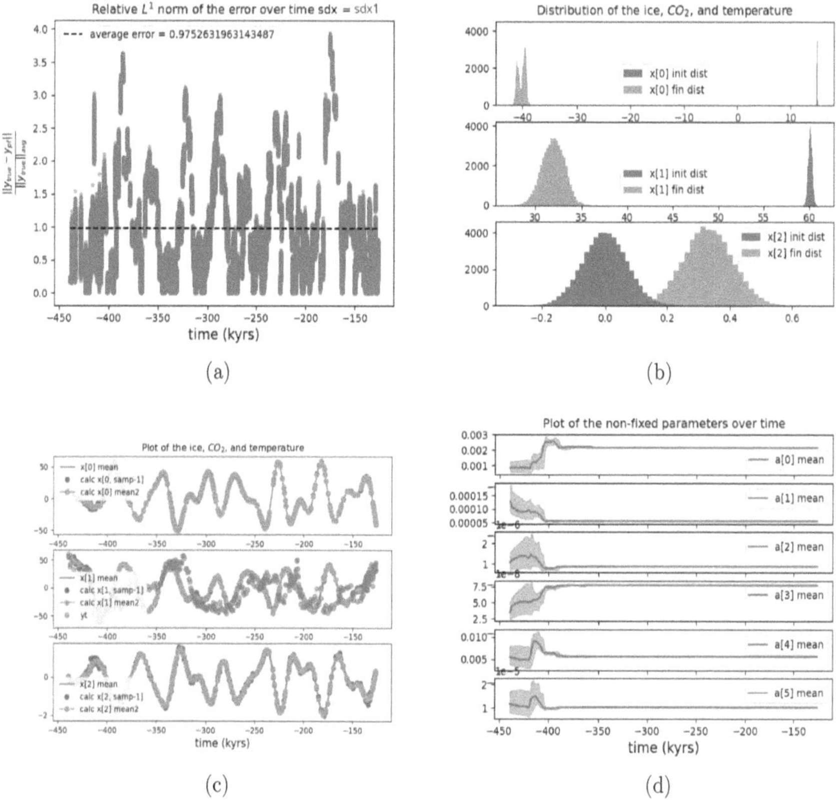

Figure 3.16: Particle filter results for $\Sigma = \mathrm{sdx1}$ and $\mathrm{nsamp} = 50{,}000$: (a) relative L^1 error vs time with the average of the particles without regard to weights (dashed), (b) distribution of state variables at initial and final time, (c) evolution of state variable distributions vs time, (d) evolution of parameter distributions vs time.

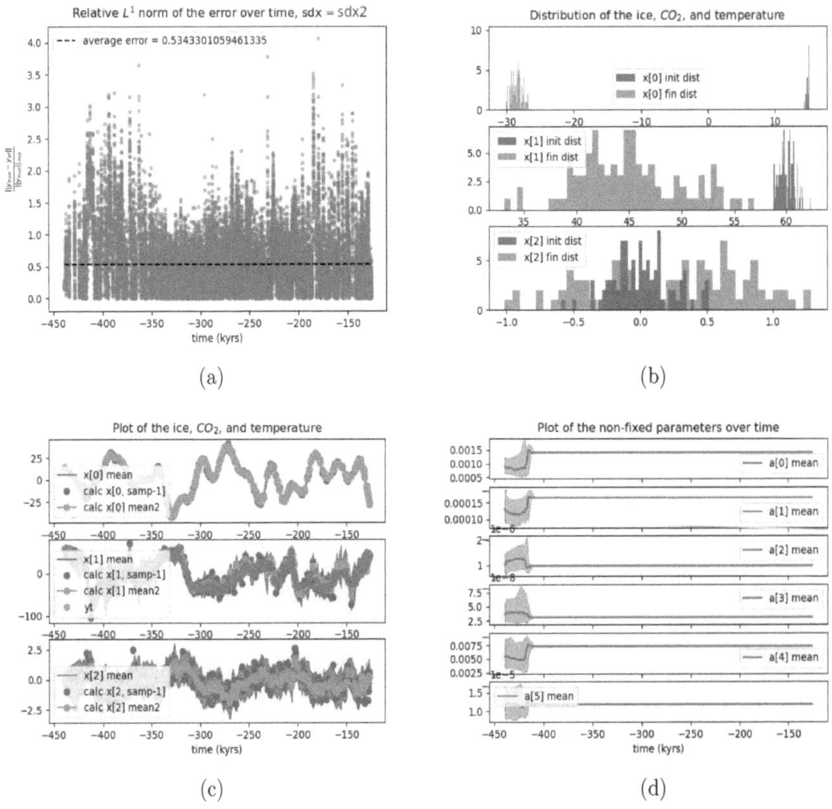

Figure 3.17: Particle filter results for $\Sigma = $ sdx2 and nsamp $= 92$: (a) relative L^1 error vs time with the average of the particles without regard to weights (dashed), (b) distribution of state variables at initial and final time, (c) evolution of state variable distributions vs time, (d) evolution of parameter distributions vs time.

Table 3.3: Average parameter values found by the particle filter using the SM91 model on the real data for sdx1 using nsamp = 50,000 and sdx2 using nsamp = 92.

param	mean (sdx1)	mean (sdx2)
a_1	2.157×10^{-3}	1.404×10^{-3}
b_1	5.601×10^{-5}	1.655×10^{-4}
b_2	8.872×10^{-7}	9.885×10^{-7}
b_3	7.760×10^{-8}	3.103×10^{-8}
b_θ	5.671×10^{-3}	7.317×10^{-3}
c_1	1.038×10^{-5}	1.194×10^{-5}

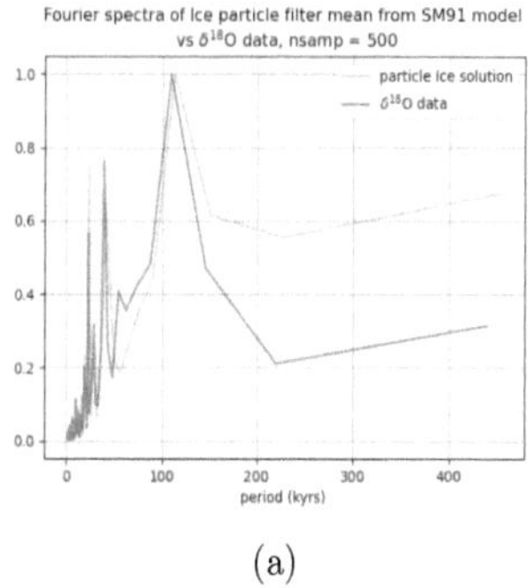

(a)

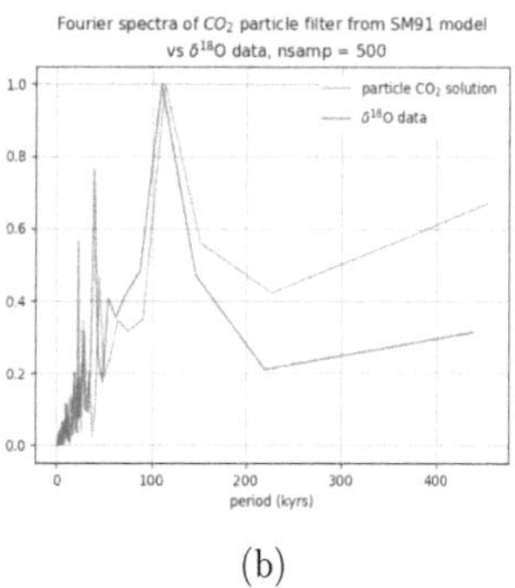

(b)

Figure 3.18: Plots of the Fourier power spectra vs period of the past 440 kyrs of the: (a) mean of the ice particle filter solution with nsamp = 500, Σ = sdx2 and δ^{18}O data [24], (b) mean of the CO_2 particle filter solution with nsamp = 500, Σ = sdx2 and δ^{18}O data [24]. Note: Fourier spectra are normalized with respect to the peak power.

Finally, the final mean parameter values for sdx1 and sdx2 are listed in Table 3.3.

As shown in Fig. 3.17 (d), the parameter distribution has collapsed to a single point, indicating that there are not enough particles to resolve the distribution, resulting in the parameter being "stuck" as a function of time, and so reducing accuracy. Seeing from the figures in the benchmarking section that as nsamp increases, the amount of variability in the parameter values also increased (see Fig. 3.4 (d) and Fig. 3.7 (d) from Section 3.1, for example). This leads us to consider increasing nsamp to, say, 500 and see if there is improvement as our data above in Fig. 3.13 (b) suggests. The last results of the particle filter for nsamp $= 500$ are shown in Fig. 3.19, and the parameter values (average) for nsamp

$= 500$ are given in Table 3.4

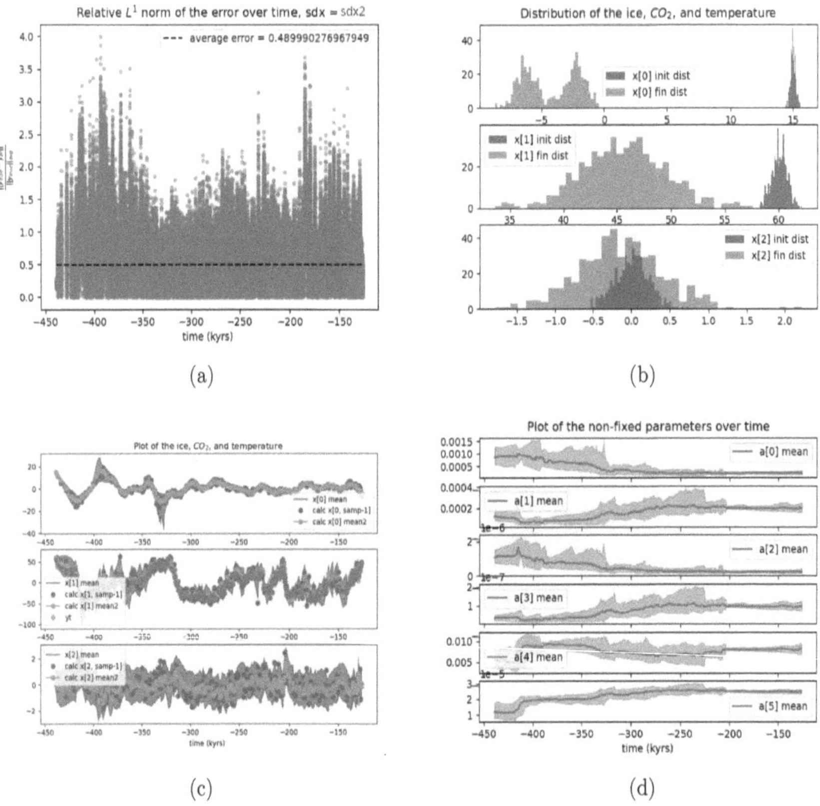

Figure 3.19: Particle filter results for Σ = sdx2 and nsamp $= 500$: (a) relative L^1 error vs time with the average of the particles without regard to weights (dashed), (b) distribution of state variables at initial and final time, (c) evolution of state variable distributions vs time, (d) evolution of parameter distributions vs time.

Thus, based on Table 3.4, each of Crucifix's parameter values (see Table 3.1) falls within one

standard deviation from these values, so there is consistency between the optimal parameter

values for the particle filter on the CO_2 data.

We now return to the "100,000 yr problem" discussed in Section 2.1. Fig. 3.18 shows

Table 3.4: Average parameter values with their standard deviations found by the particle filter using the SM91 model using sdx2 and nsamp = 500 on the real data.

param	mean	standard deviation
a_1	1.011×10^{-3}	7.413×10^{-4}
b_1	9.413×10^{-5}	3.618×10^{-5}
b_2	1.158×10^{-6}	5.660×10^{-7}
b_3	4.540×10^{-8}	1.832×10^{-8}
b_θ	6.670×10^{-3}	2.614×10^{-3}
c_1	1.115×10^{-5}	5.689×10^{-6}

a comparison of the Fourier spectra from our calculated solutions and the measurements for ice (Fig. 3.18(a)) and CO_2 (Fig. 3.18(b)). The SM91 model with Milakovitch forcing is able to reproduce the strong signal in the data at 100 kyr that could not be described using Milankovitch forcing alone.

3.3 Model selection using real data

In Section 3.2 we have successfully implemented the particle filter to determine the states and parameters for the SM91 model following the ideas in Crucifix & Rougier (2009) [4]. In Crucifix & Rougier (2009) the authors state that "one must hope that an extra validation will lead us to prefer one model to the other. The contrary will be an indication that model discrepancy was underestimated" (p. 28). In a later paper [3] Crucifix states that subsequent unpublished results on the use of the particle filter for model selection were not encouraging (p. 839):

"... further sensitivity studies performed after publication lead us to tone down our enthusiasm. The filter performs poorly on the model selection problem because it fails to discriminate models on the basis of their longterm dynamics."

Here, we will look into the claim made by Crucifix ([3], p. 839) that the particle filter
wasn't able to "select" the "correct" model based on the given data. First, we'll look at how
the particle filter (using the Vostok CO_2 data) reacts to the Lorenz model ([13], p. 135) and
see if it tells us that the model is a bad choice (as expected). Then, we will move on to
other Saltzmann-type models used for modelling the ice, CO_2 and ocean temperature. We
conclude that the particle filter does appear to be "selecting" an optimal model based on
the CO_2 data alone.

Lorenz Model

The Lorenz system is a set of convection equations which are derived by taking truncated
approximations of a Fourier series solution in an atmospheric convection model ([13], p.
134 - 135). The system was used as a simple example that would illustrate deterministic
nonperiodic flow ([13], p. 134). Again, for us, this model is chosen as an example of a "bad"
model for ice ages which has the same number of variables as the SM91 model.

This system can be written as:

$$
\begin{aligned}
\frac{dx}{dt} &= \sigma \left(y - x \right) \\
\frac{dy}{dt} &= x \left(\rho - z \right) - y \\
\frac{dz}{dt} &= xy - \beta z
\end{aligned}
\tag{3.5}
$$

where x, y, and z are the state variables and σ, ρ, and β are the system parameters. Now,
with respect to SM91 model, it is not clear how x, y or z would correspond to I, μ or θ (or
how to translate the Lorenz parameters to SM91 for that matter) but to see if the particle

filter will recognize that the Lorenz model is a bad choice, we will rewrite Eq. (3.5) in terms of the varaiables and parameters of the SM91 model by choosing $x = I$, $y = \mu$, and $z = \theta$:

$$\frac{dI}{dt} = \sigma \left(\mu - I \right)$$

$$\frac{d\mu}{dt} = I \left(\rho - \theta \right) - \mu \tag{3.6}$$

$$\frac{d\theta}{dt} = I\mu - \beta\theta$$

For this system, we will initialize the parameters to be: $\sigma = 8.7 \times 10^{-4}$, $\rho = 1.3 \times 10^{-4}$ and $\beta = 1.1 \times 10^{-6}$ (using the same lognormal distribution as in SM91); the initial conditions will be $x_0 = 15.0$, $y_0 \approx 51.0$, and $z_0 = 0.0$. Looking at the code in Appendix A, lines 141 - 149 would change to:

```
1   # The Lorenz system:
2   a1 = A[0]; b1 = A[1]; b2 = A[2]
3   z[0] = -a1*I + a1*mu
4   z[1] = -I*theta + b1*I - mu
5   z[2] =  I*mu - b2*theta
```

Running the code with sdy = 5.0 and rs_thresh = 1.0 as in the previous sections with these modifications, the algorithm became unstable and would give warnings such as:

```
RuntimeWarning:  invalid value encountered in add
```

```
x[i+1,:,k] = x[i,:,k] + dt[i]*dfdt(tau[i], x[i,:,k], phi, R[i])
```

```
+ sdx*dW[i,:,k]
```

or errors like:

```
SVD did not converge
```

```
Error in the weighted average.
```

Because the algorithm failed, no relative error data could be collected.

Thus, it appears so far that the algorithm correctly finds this model to not be suitable for the

CO_2 data. It might be possible that some other permuation of these variables in Eq. (3.6)

may make it not fail, however, but the point is that a poor model choice should result in

poor quality in the output from the particle filter.

Saltzman and Maasch 1990 model

The Saltzman and Maasch 1990 model [15], p. 1955 (from now on referred to as SM90)

is given by:

$$\frac{dI}{dt} = -a_0 I - a_1 \mu - a_2 M(t)$$

$$\frac{d\mu}{dt} = b_1 \mu - (b_2 - b_3 N)N - b_4 N^2 \mu \tag{3.7}$$

$$\frac{dN}{dt} = -c_0 I - c_2 N$$

Some of the main differences between SM90 and SM91 are the eqations for dI/dt and $d\mu/dt$,

particularly the nonlinear terms in the $d\mu/dt$ equation. Another difference is that N in

Eq. (3.7) represents the "deep ocean warmth and salinity" ([15], p. 1955), as opposed to

θ being the ocean temperature anomaly. Also, M(t) in Eq. (3.7) is the same as R(t), the

external insolation forcing. Since the particle filter appears to run successfully using the

SM91 model in Section 3.2, we decided to rearrange the SM91 model as much as we can to

make it "look like" the SM90 model by first setting $\theta = N$ and then matching terms that

are the same:

$$a_0 = K_I$$

$$(a_1)_{SM90} = (a_1)_{SM91} k_\mu$$

$$k_\theta = 0$$

$$a_2 = (a_1)_{SM91} K_R$$

$$(b_1)_{SM90} = (b_1)_{SM91} \tag{3.8}$$

$$(b_2)_{SM90} = b_\theta$$

$$(b_3)_{SM90} = -(b_2)_{SM91}$$

$$b_4 = (b_3)_{SM91}$$

$$c_0 = c_1$$

$$c_2 = K_\theta$$

where the SM90 or SM91 subscripts refer to the model that those coefficients pertain to in order to prevent confusion. Then translation between models is complete if we replace $-b_2\mu^2 - b_3\mu^3$ in SM91 with $b_3\theta^2 - b_4\theta^2\mu$ in SM90. Hence lines 141 - 149 of the code should look like

```
1  # The SM90 model:
2
3  ktheta = 0.
4  a1 = A[0]; b1 = A[1]; b2 = A[2]
5  b3 = A[3]; btheta = A[4]; c1 = A[5];
6
7  z[0] = -KI*I -(a1*kmu)*mu - (a1*ktheta)*theta - (a1*kR)*R
8  z[1] =   b1*mu - btheta*theta + b2*theta**2 - b3*theta**2*mu
9  z[2] = -c1*I - Ktheta*theta
```

In Appendix D we conduct a benchmarking error analysis of the SM90 model in terms of the particle filter computational parameters by applying the algorithm to the Vostok CO_2 data. The SM90 model yielded promising results with sdx2 giving better stability and error profile than sdx1 (see Appendix D).

We now use the particle filter as in Section 3.2 to fit the SM90 model to the Vostok CO_2 data. The central values of the non-fixed parameters from Table 1 on p. 24 of Crucifix & Rougier (2009) (see our Table 3.1) were mostly chosen to be the same for the corresponding parameters in SM90 model except for $k_\theta = 0$. All parameters of this model will also have a lognormal distribution as used for the SM91 model. As stated in Appendix D, we will choose sdy = 10.0 and rs_thresh = 0.7 for sdx1, and sdy = 5.0 and rs_thresh = 1.0 for sdx2. Comparing the E profiles of sdx1 and sdx2, we can see that there is a significant difference between the two curves (see Fig. 3.20). From Fig. 3.21, the data for sdx2 fits an exponential curve just like in Section 3.2 with an asymptote of about 0.272 (compared to 0.248 for sdx2 using the SM91 model).

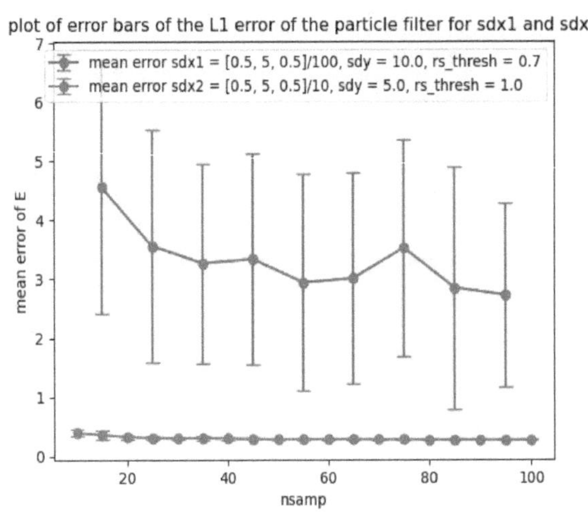

Figure 3.20: Plots of the mean of E for the SM90 model using both sdx1 and sdx2 versus the number of samples.

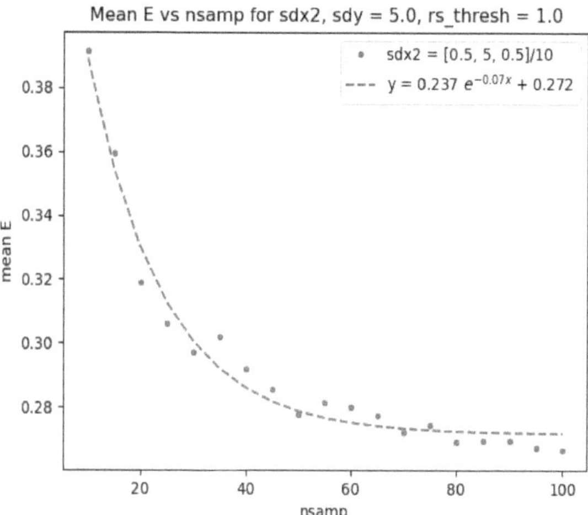

Figure 3.21: An exponential model of the mean of E for the SM90 model using sdx2 versus the number of samples.

As was done in the previous section, we will give an in-depth breakdown of the distribution of the particle filter and parameter data, including how the error, parameters, and state variables vary over time. Please note that the parameters that will follow are the non-fixed parameters that would correspond to the ones in the SM91 model. Doing the analysis for sdx1 with sdy $= 10.0$, rs_thresh $= 0.7$, and nsamp $= 100$, one can see how poorly the particle filter did for sdx1 by looking at Fig. 3.22 (a) and (c), which may explain the large E value in Fig. 3.20 for sdx1. Repeating the plots for sdx2, Fig. 3.23 (a) and (c) tells us that the algorithm did very well in following the measured CO_2 data; they are comparable to Fig. 3.17 (a) and (c) for the SM91 model.

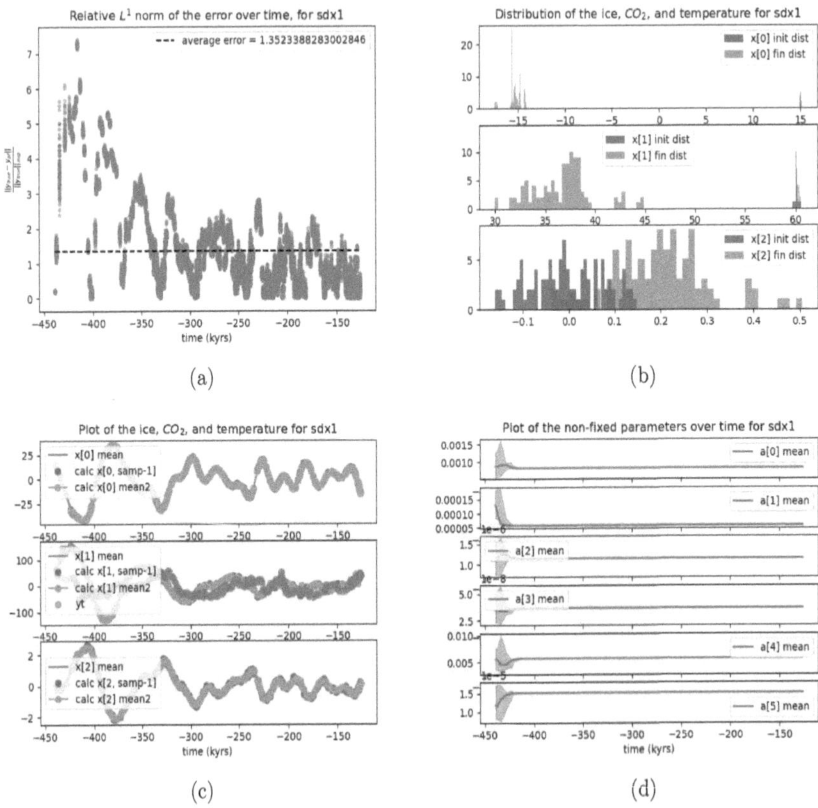

Figure 3.22: Particle filter results for the SM90 model with $\Sigma = \text{sdx1}$ and $\text{nsamp} = 100$: (a) relative L^1 error vs time with the average of the particles without regard to weights (dashed), (b) distribution of state variables at initial and final time, (c) evolution of state variable distributions vs time, (d) evolution of parameter distributions vs time.

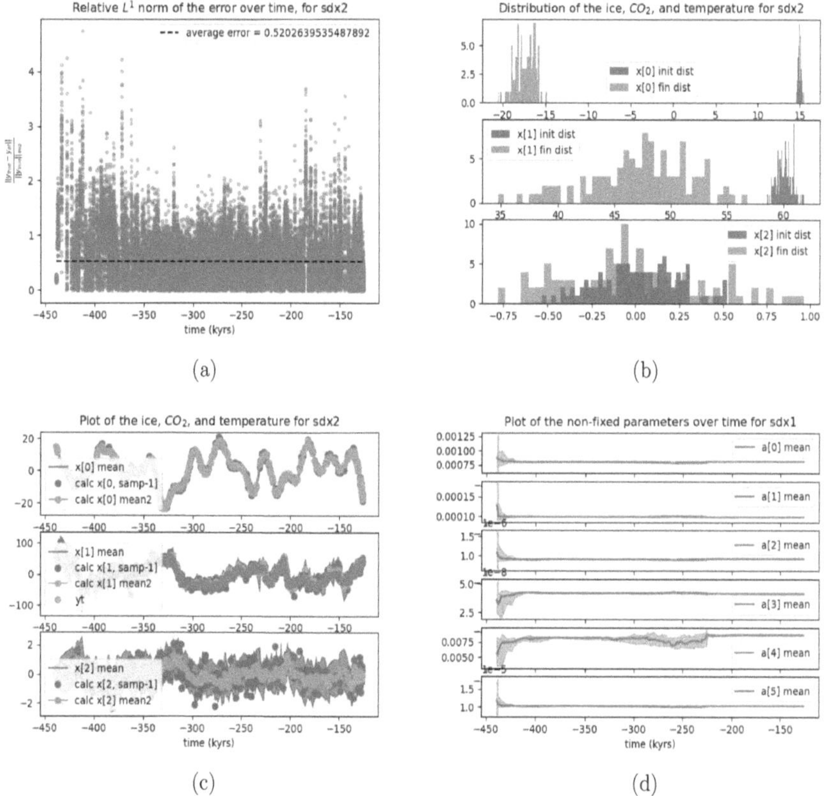

Figure 3.23: Particle filter results for for the SM90 model $\Sigma = $ sdx2 and nsamp $= 100$: (a) relative L^1 error vs time with the average of the particles without regard to weights (dashed), (b) distribution of state variables at initial and final time, (c) evolution of state variable distributions vs time, (d) evolution of parameter distributions vs time.

Now lets move on to a slight modification to Eq. (3.7) (from now on will be labeled as BR22):

$$\frac{dI}{dt} = -a_0 I - a_1 \mu - a_2 R(t)$$

$$\frac{d\mu}{dt} = b_1 \mu - (b_2 - b_3 \theta)\theta - 2b_4 \theta \mu \qquad (3.9)$$

$$\frac{d\theta}{dt} = -c_0 I - c_2 \theta$$

where the only difference from the SM90 is in the second equation, where the $\theta^2\mu$ term in

the SM90 is changed to $2\theta\mu$. Lines 141 - 149 of the code should look like

```
# BR22 # similar to sm90 but with a quadratic term in the second equation
    instead of cubic

ktheta = 0.
a1 = A[0];  b1 = A[1];  b2 = 0.
b3 = A[3];  btheta = A[4];  c1 = A[5];

z[0] = -KI*I -(a1*kmu)*mu - (a1*ktheta)*theta - (a1*kR)*R
z[1] =   b1*mu - btheta*theta + b2*theta**2 - 2*b3*theta*mu
z[2] = -c1*I - Ktheta*theta
```

We will choose sdy $= 10.0$ and rs_thresh $= 0.7$ for sdx1, and sdy $= 5.0$ and rs_thresh $=$

1.0 for sdx2 (see Appendix E for justification). We now apply the particle filter method to

the Vostok CO_2 data for the BR22 model. The results are shown in Fig. 3.24 for sdx1 and

Fig. 3.25 for sdx2. Once again, the results were better for sdx2 than sdx1 as expected. What

is interesting though is that the average relative error as shown in Fig. 3.25(a) for BR22 is

slightly lower than the average relative error in Fig. 3.17(a) and Fig. 3.19(a) for SM91. But,

there is a lack of variability in the parameters over time as depicted in Figures 3.17(d) and

3.25(d). However, the exponential model for E using sdx2 as shown in Fig. 3.26 for BR22

is not lower than the one for sdx2 shown in Fig. 3.14 for SM91. This may suggest that the

SM91 model using sdx2 may still be a better candidate than these other models so far.

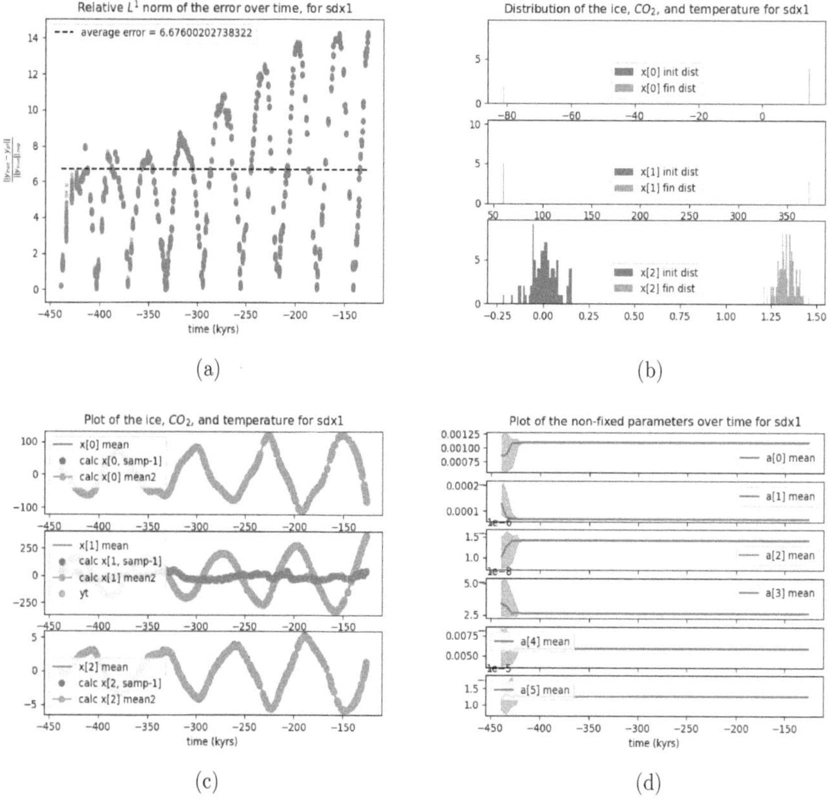

Figure 3.24: Particle filter results for for the BR22 model $\Sigma = \mathrm{sdx1}$ and $\mathrm{nsamp} = 100$: (a) relative L^1 error vs time with the average of the particles without regard to weights (dashed), (b) distribution of state variables at initial and final time, (c) evolution of state variable distributions vs time, (d) evolution of parameter distributions vs time.

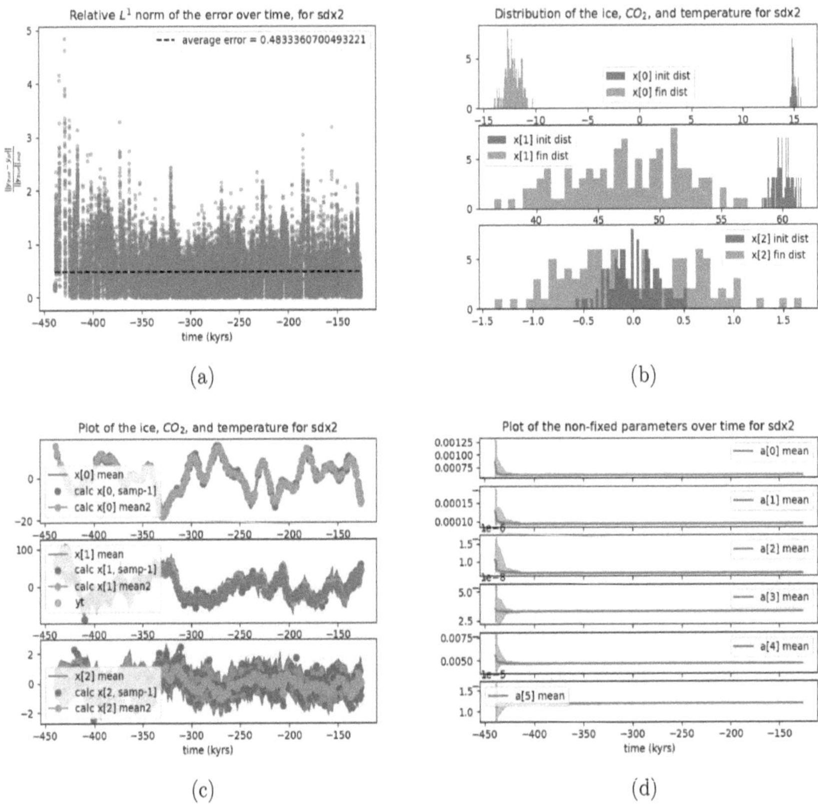

Figure 3.25: Particle filter results for for the BR22 model $\Sigma = \text{sdx2}$ and $\text{nsamp} = 100$: (a) relative L^1 error vs time with the average of the particles without regard to weights (dashed), (b) distribution of state variables at initial and final time, (c) evolution of state variable distributions vs time, (d) evolution of parameter distributions vs time.

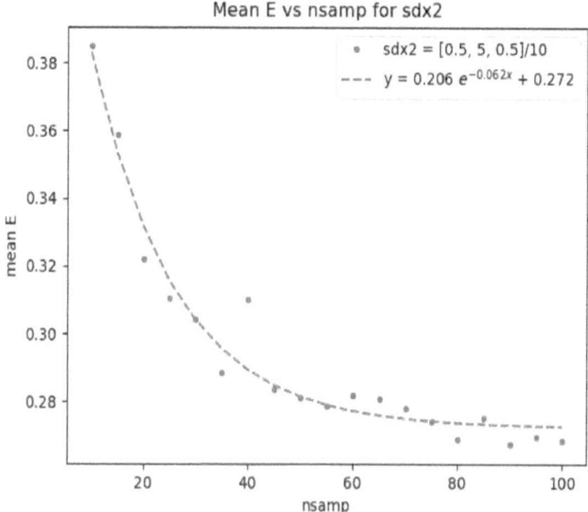

Figure 3.26: An exponential model of the mean of E for sdx2 for the BR22 model.

Saltzman and Sutera 1987 model

A fourth model that we'll consider is the Saltzman and Sutera 1987 model [20] (SS87). The reduced model with no forcing is

$$\frac{dI}{dt} = -a_0 I - a_1 \mu$$

$$\frac{d\mu}{dt} = b_1 \mu + b_5 \theta - b_6 \theta^2 \mu \qquad (3.10)$$

$$\frac{d\theta}{dt} = c_0 I - c_2 \theta$$

([20], p. 238) which is strikingly similar to the SM90 model. Comparing SM90 model with the SS87 model, we would only need to make a few minor adjustments:

$$(a_2)_{SM90} = 0$$

$$(b_5)_{SS87} = -(b_2)_{SM90}$$

$$(b_3)_{SM90} = 0 \tag{3.11}$$

$$(b_6)_{SS87} = (b_4)_{SM90}$$

$$(c_0)_{SS87} = -(c_0)_{SM90}$$

Hence lines 141 - 149 of the code should look like

```
1 # Saltzman and Sutera 1987
2 ktheta = 0.; kR = 0.
3 a1 = A[0]; b1 = A[1]; b2 = 0.
4 b3 = A[3]; btheta = A[4]; c1 = A[5];
5
6 z[0] = -KI*I -(a1*kmu)*mu - (a1*ktheta)*theta - (a1*kR)*R
7 z[1] =   b1*mu + btheta*theta + b2*theta**2 - b3*theta**2*mu
8 z[2] = c1*I - Ktheta*theta
```

In Appendix F we provide the benchmarking analysis for the SS87 model using the relative error data for the CO_2. As determined in Appendix F, we will use sdy = 10.0 and rs_thresh = 0.8 for sdx1, and sdy = 5.0 and rs_thresh = 1.0 for sdx2 with this model. Fig. 3.27 gives the E profile for sdx2 and it is slightly lower than the E profile for BR22 in Fig. 3.26. The summary analysis of the particle filter distributions for states and parameters versus time is given in Fig. 3.28 for sdx1 and Fig. 3.29 for sdx2. Comparing Fig. 3.28(c) and 3.29(c), we can see that the SS87 model using sdx2 tracks the data closely while that for sdx1 does poorly. Looking at Fig. 3.29(a), the average relative error is lower than even the one in Fig. 3.19(a) for SM91 as well. But, as shown in Fig. 3.27, the asymptote of its E profile is not any lower than the one for SM91 in Fig. 3.14.

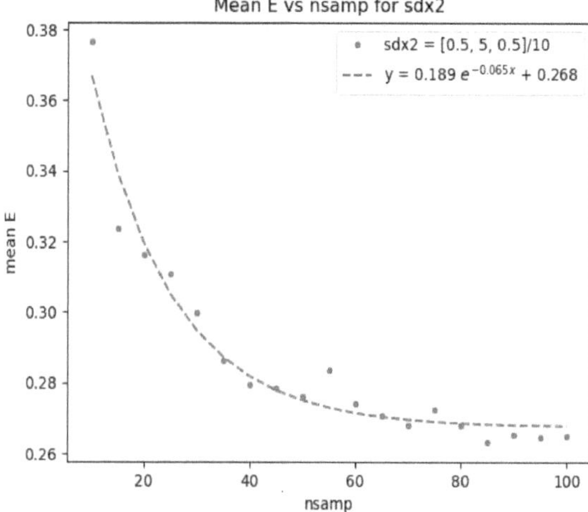

Figure 3.27: An exponential model of the mean of E for sdx2 for the SS87 model.

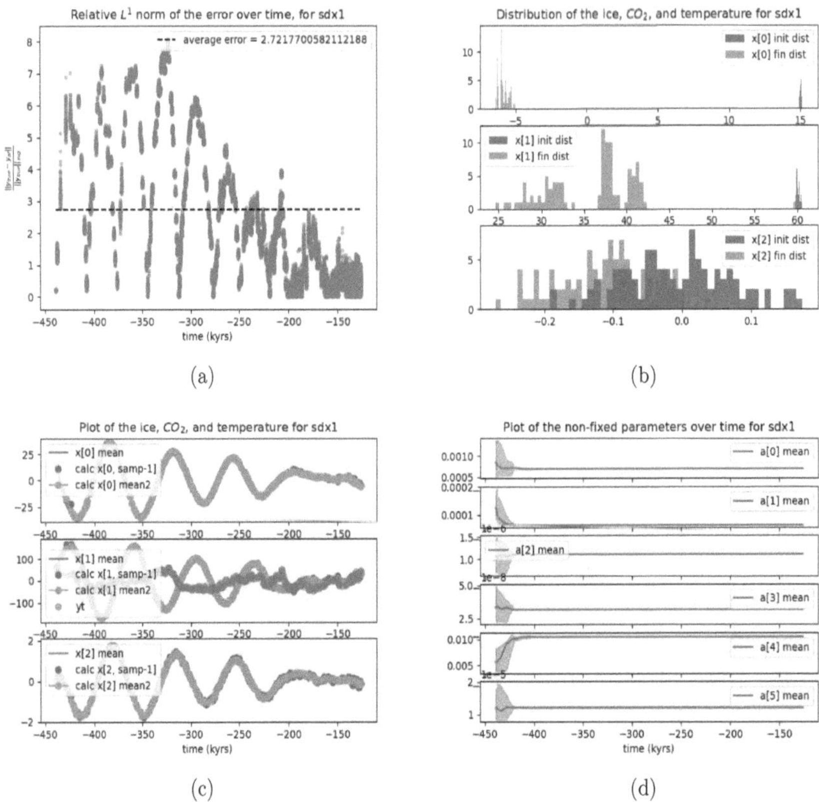

Figure 3.28: Particle filter results for for the SS87 model $\Sigma = $ sdx1 and nsamp $= 100$: (a) relative L^1 error vs time with the average of the particles without regard to weights (dashed), (b) distribution of state variables at initial and final time, (c) evolution of state variable distributions vs time, (d) evolution of parameter distributions vs time.

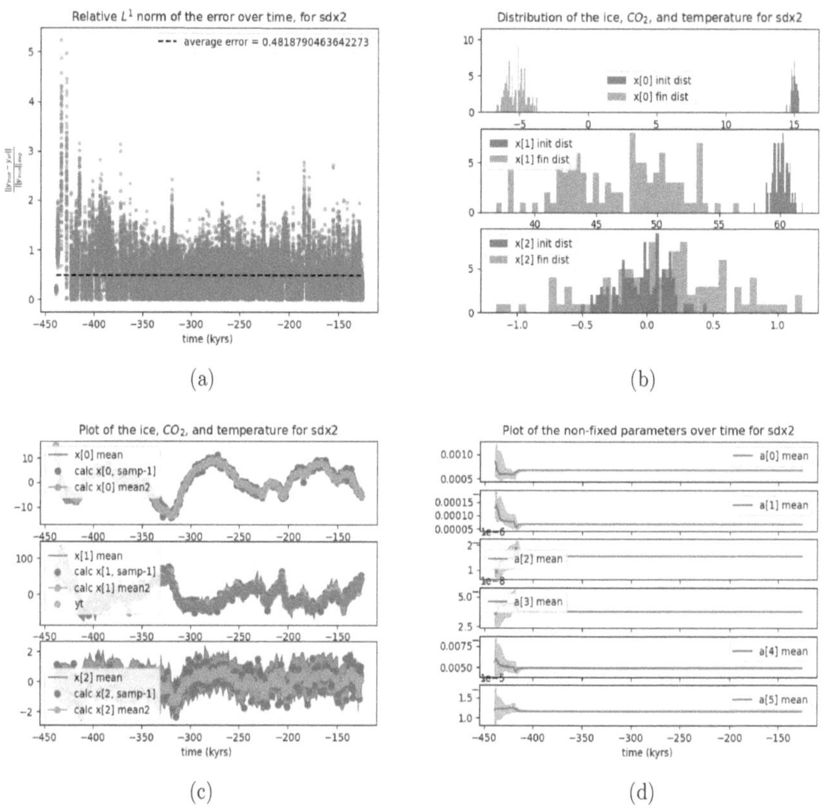

Figure 3.29: Particle filter results for for the SS87 model Σ = sdx2 and nsamp = 100: (a) relative L^1 error vs time with the average of the particles without regard to weights (dashed), (b) distribution of state variables at initial and final time, (c) evolution of state variable distributions vs time, (d) evolution of parameter distributions vs time.

Discussion of results

Based on the analysis done above, it does appear that the particle filter algorithm is, in a sense, selecting the "better" model over another. The Lorenz model for example completely failed to give results (even though it is a trivial case as it is not clear as to which variable is suitable for ice, CO_2, or ocean temperature, let alone the meaning of its parameters). The results did seem to favor models that resemble the SM91 model in some way. The differences in the E profiles in the SS87, SM90, BR22, and SM91 are fairly close when using sdx2, but the asymptote of the exponential model for SM91 is the lowest of them all (one could argue that it is not significantly lower than the others, but it is the lowest nonetheless). One explanation for their E profiles being similar is the simple fact that each model is similar to one another, up to some nonlinear terms. Also notice that sometimes the average relative error for one model is lower than another but its E profile is not any lower compared to the rest of the models. One possible reason for this despite having a not-so-low E profile could be because that average relative error is only one run of the particle filter while the E profile considers several runs over time. But it's intriguing that there are similarities in the CO_2 graphs using sdx2 regardless if there was forcing or no forcing. This leads me to consider one more case where we use the SM91 model as in the last section except without forcing – will removing the external forcing cause any significant changes, perhaps a worse E profile?

Unforced SM91 model

We thus revisit the SM91 model but explicitly set $k_R = 0$ to remove the orbital forcing. The results in Appendix G explains our choice of rs_thresh $= 1.0$ and sdy $= 5.0$.

The average relative error shown in Figures 3.30(a) for sdx1 and 3.31(a) for sdx2 are both lower than for all of the other models in this section (when comparing to their respective sdx1 and sdx2 results) – even the average relative error is lower than the one in Fig. 3.19(a) for sdx2 (bear in mind that this is only for one trial, not multiple trials like in any of the E profile results). As depicted in Fig. 3.31(c), the algorithm fitted this model well with the

CO_2 data. The E profiles for sdx1 and sdx2 shown in Fig. 3.32(a) exhibited good results as well, but the exponential asymptote for the E profile for sdx2 with no external forcing shown in Fig. 3.32(b) is not lower than the one with external forcing as shown in Fig. 3.14.

Summary of the model comparison

Our comparison of different ice age models using parameter selection via the particle filter method was based on filtering the CO_2 predictions to the Vostok CO_2 data. We found that the "bad" model did not work and the "good" models were all reasonably competitive, and identification of the "best" model was not clear cut. At this point the selection problem falls into the category described by Crucifix & Rougier (2009) on p. 28,

> ... an issue may arise if structurally different models equally pass the validation test but yield incompatible predictions. In this case one must hope that an extra validation criteria will lead us to prefer one model to the other.

In the next section we consider such an additional validation criteria, that of the ice volume of the glaciers.

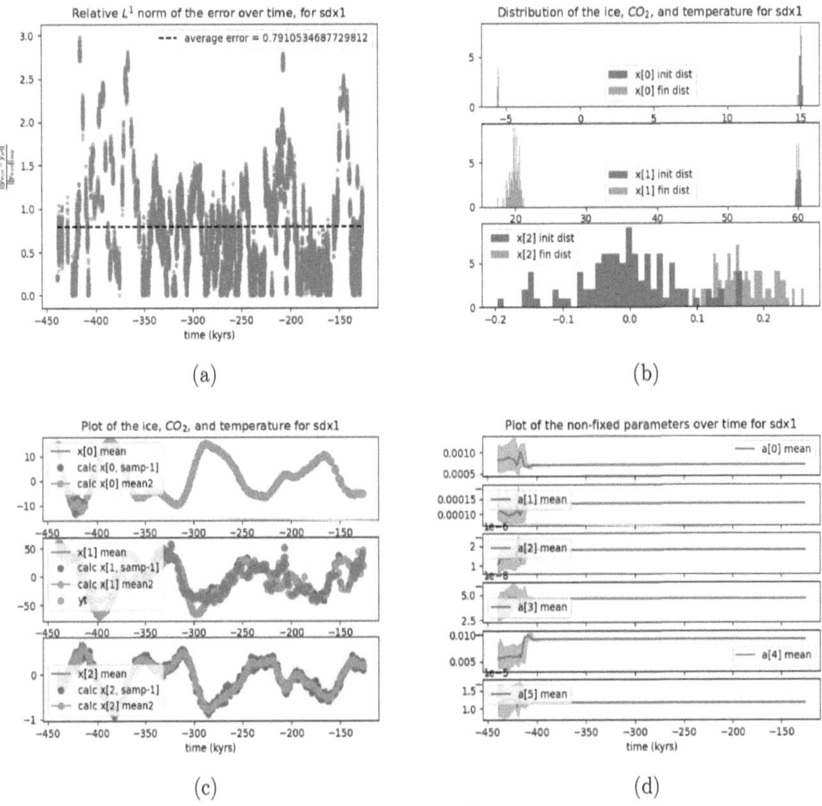

Figure 3.30: Particle filter results for for the unforced SM91 model $\Sigma = \text{sdx1}$ and nsamp $=$ 100: (a) relative L^1 error vs time with the average of the particles without regard to weights (dashed), (b) distribution of state variables at initial and final time, (c) evolution of state variable distributions vs time, (d) evolution of parameter distributions vs time.

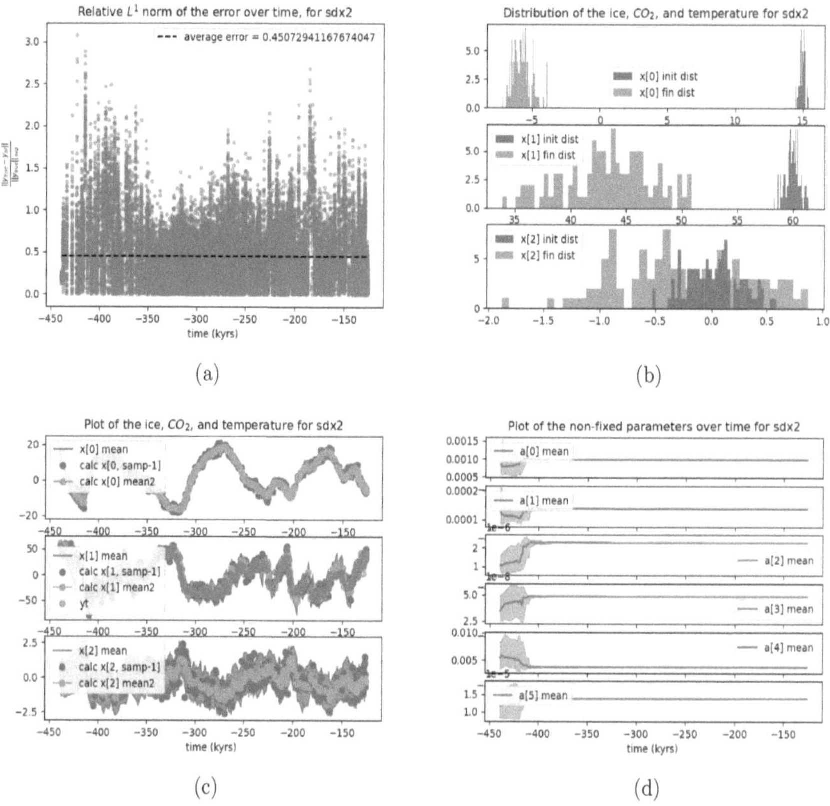

Figure 3.31: Particle filter results for for the unforced SM91 model $\Sigma = \text{sdx2}$ and nsamp = 100: (a) relative L^1 error vs time with the average of the particles without regard to weights (dashed), (b) distribution of state variables at initial and final time, (c) evolution of state variable distributions vs time, (d) evolution of parameter distributions vs time.

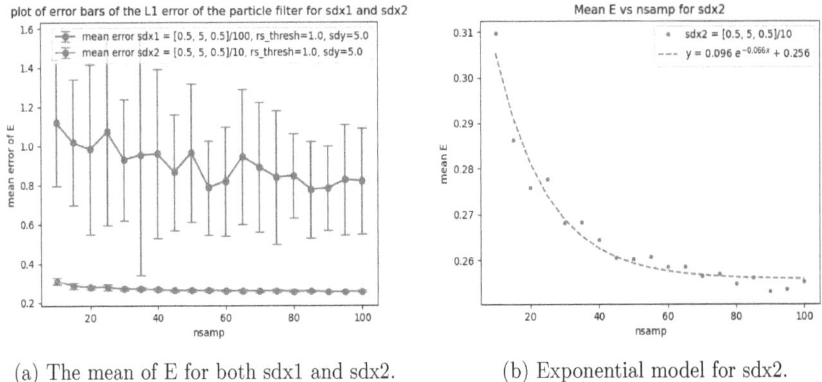

(a) The mean of E for both sdx1 and sdx2.   (b) Exponential model for sdx2.

Figure 3.32: The average E data for the SM91 model without forcing for various number of samples.

3.4   Model selection using ice volume data proxy

In Section 3.3, we saw that the particle filter seems not to be favoring one model over another. Here, we will further verify whether or not this is the case by including the SPECMAP data [7] as a proxy for the ice data into the error analysis as was done in Crucifix & Rougier (2009). Using our previous results from the particle filter based on filtering the CO_2 data we will calculate the relative error of the model in predicting the ice volume. Since only the CO_2 measurements are used in the particle filter for the determination of the model states and parameters, prediction of ice volume represents a more sensitive test of model validity (in contrast to the prediction of CO_2 measurements which are used directly by the particle method).

As is noted in Crucifix & Rougier (2009), p. 25, the SPECMAP data is not a pure proxy for ice volume as it is also somewhat affected by temperature. Following Crucifix & Rougier

76

(2009), to compare to the ice volume predictions of the particle filter we scale the SPECMAP
data based on the mean and variance of the ice volume obtained from the particle filter. This
will be done as follows: suppose the SPECMAP time series data relative to its mean is

$$S = \{s_1, s_2, ..., s_N\} \tag{3.12}$$

where N is the number of data points and the ice data (relative to its mean) from the particle
filter is

$$I = \{i_1, i_2, ..., i_N\} \tag{3.13}$$

We introduce a scaled SPECMAP data as

$$\tilde{S} = \{\alpha s_1, \alpha s_2, ..., \alpha s_N\} = \alpha S \tag{3.14}$$

where α is some positive real number. Since

$$S_{avg} = \frac{1}{N} \sum_{k=1}^{N} s_k = 0 \tag{3.15}$$

$$I_{avg} = \frac{1}{N} \sum_{k=1}^{N} i_k = 0 , \tag{3.16}$$

it follows that

$$S_{var} = \frac{1}{N}\sum_{k=1}^{N} s_k^2 \tag{3.17}$$

$$\tilde{S}_{avg} = \frac{\alpha}{N}\sum_{k=1}^{N} s_k = 0 \tag{3.18}$$

$$\tilde{S}_{var} = \frac{\alpha^2}{N}\sum_{k=1}^{N} s_k^2 = \alpha^2 S_{var} \tag{3.19}$$

$$I_{var} = \frac{1}{N}\sum_{k=1}^{N} i_k^2 \tag{3.20}$$

If we let the variance of the scaled SPECMAP to be equal to the variance of the particle filter ice data, we can then get α from

$$\tilde{S}_{var} = I_{var} \tag{3.21}$$

$$\alpha^2 S_{var} = I_{var} \tag{3.22}$$

$$\alpha = \sqrt{\frac{I_{var}}{S_{var}}} \tag{3.23}$$

We are now ready to redo the error analysis with the scaled SPECMAP. Throughout this section, we will use nsamp = 500, sdy = 5, Σ = sdx2 and rs_thresh = 1.0. Starting with SM91 model, here are the results for the insolation forcing case:

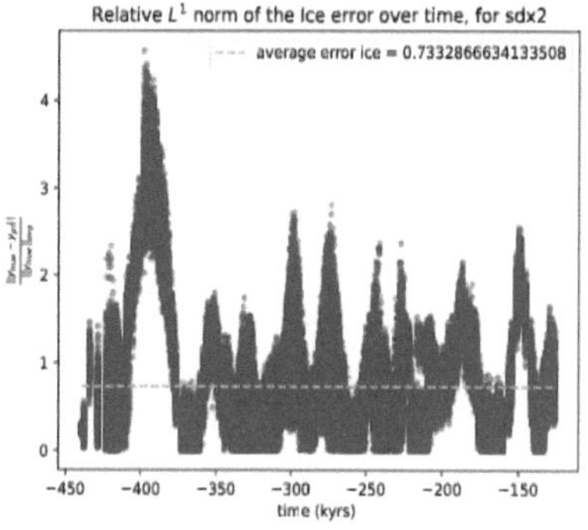

(a) The ice relative error over time.

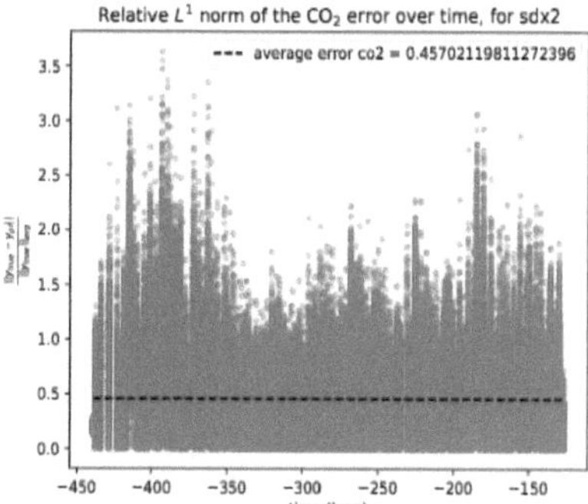

(b) The CO_2 relative error over time.

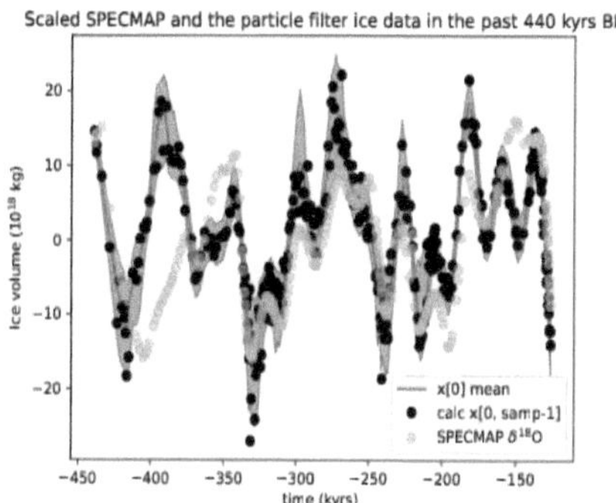

(c) The ice data over time.

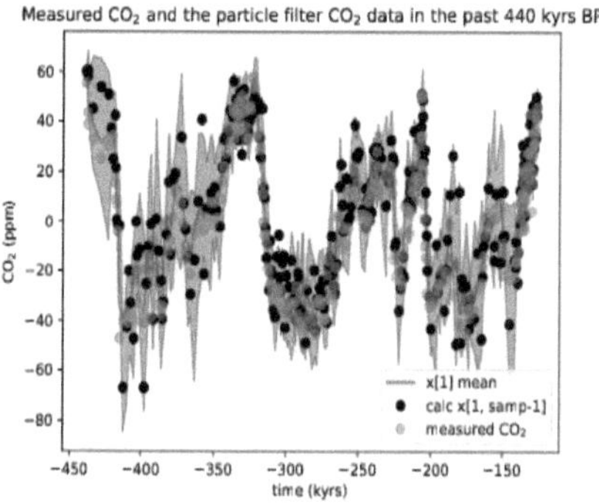

(d) The CO_2 data over time.

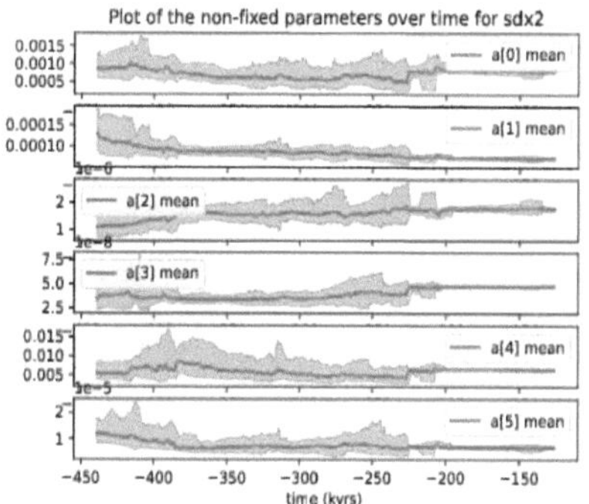

(e) The parameter distribution over time.

Figure 3.33: Results for the SM91 model with insolation forcing.

and the results without insolation forcing:

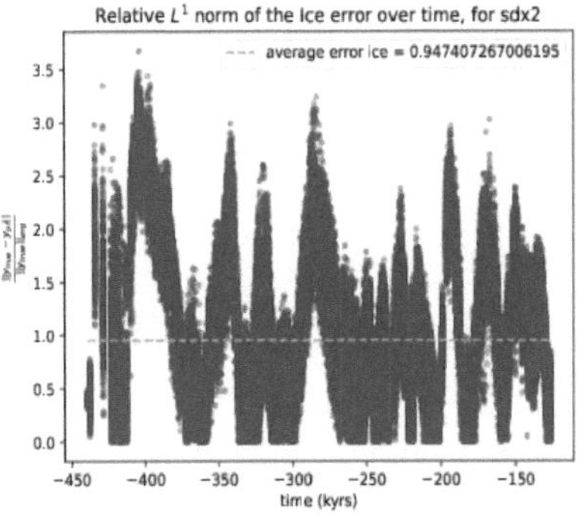

(a) The ice relative error over time.

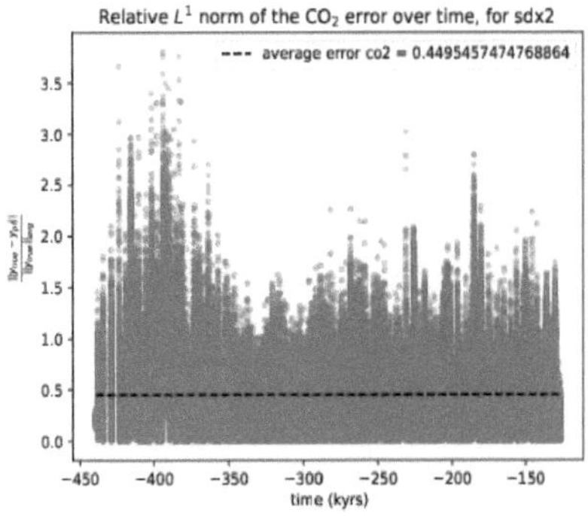

(b) The CO_2 relative error over time.

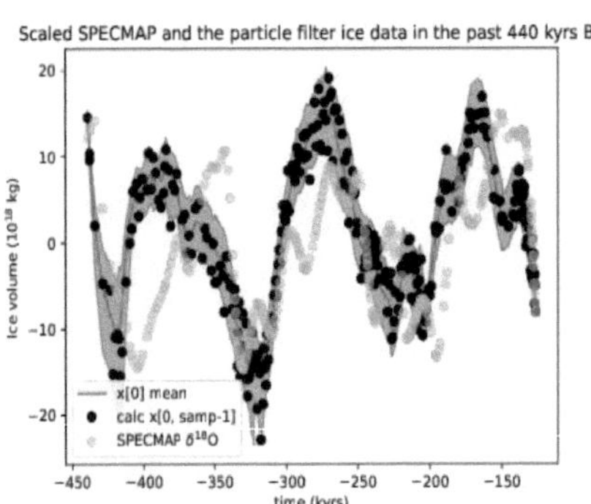

(c) The ice data over time.

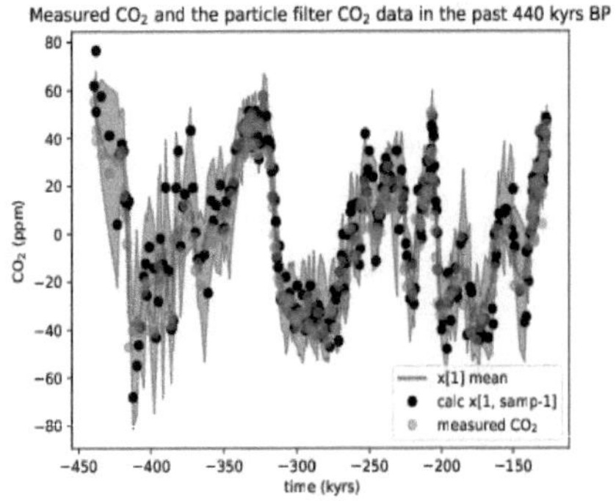

(d) The CO_2 data over time.

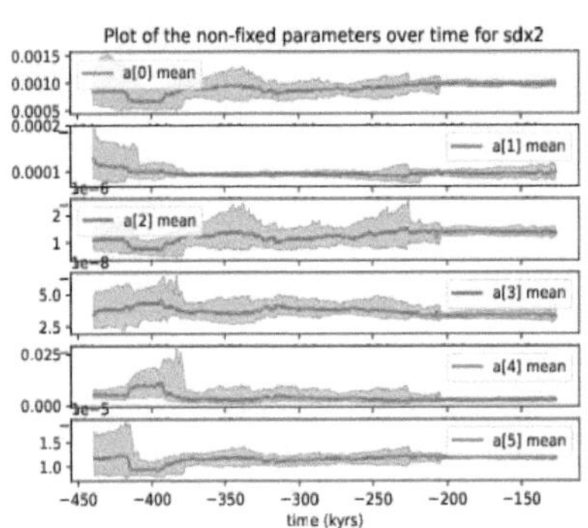

(e) The parameter distribution over time.

Figure 3.34: Results for the SM91 model without insolation forcing.

Next, for the SM90 model, here are the results for the insolation forcing case:

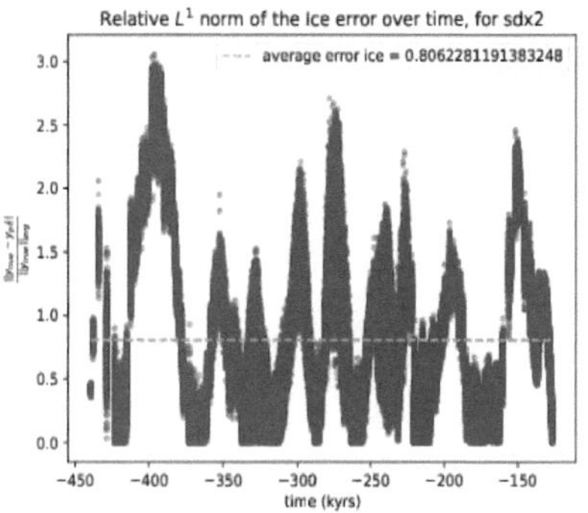

(a) The ice relative error over time.

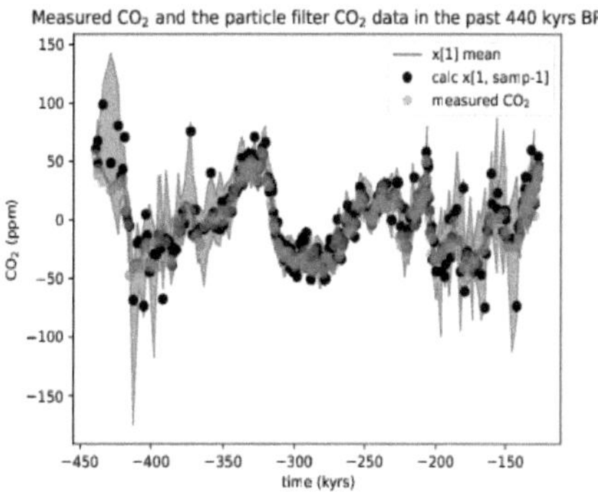

(b) The CO_2 relative error over time.

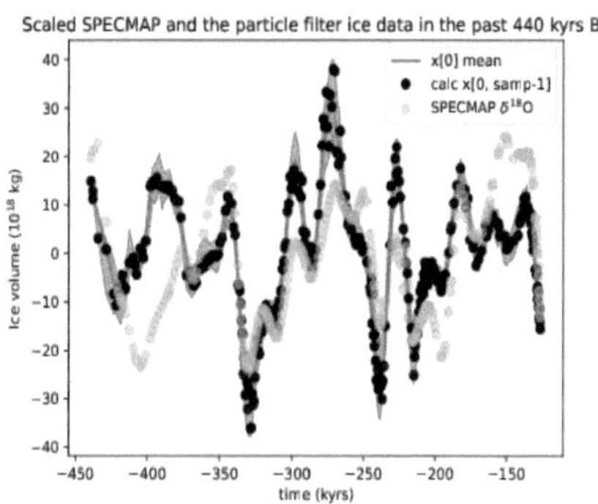

(c) The ice data over time.

(d) The CO_2 data over time.

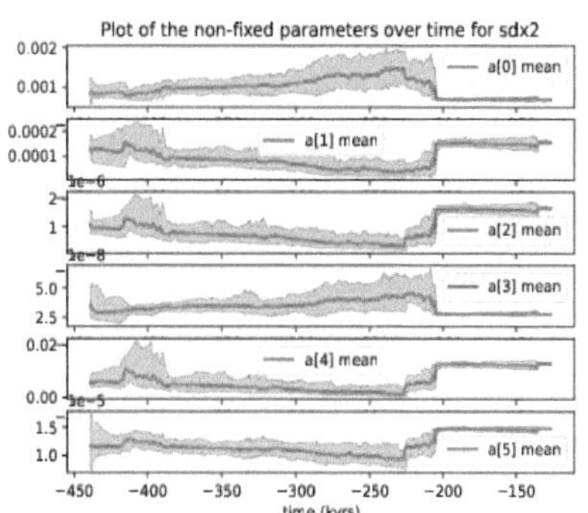

(e) The parameter distribution over time.

Figure 3.35: Results for the SM90 model with insolation forcing.

and the results for the SM90 model without insolation forcing:

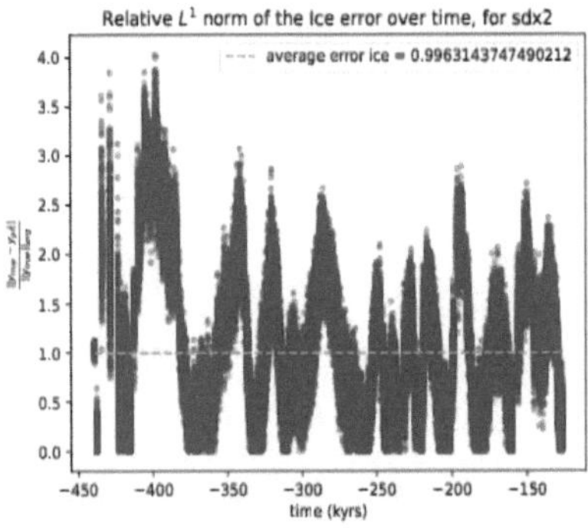

(a) The ice relative error over time.

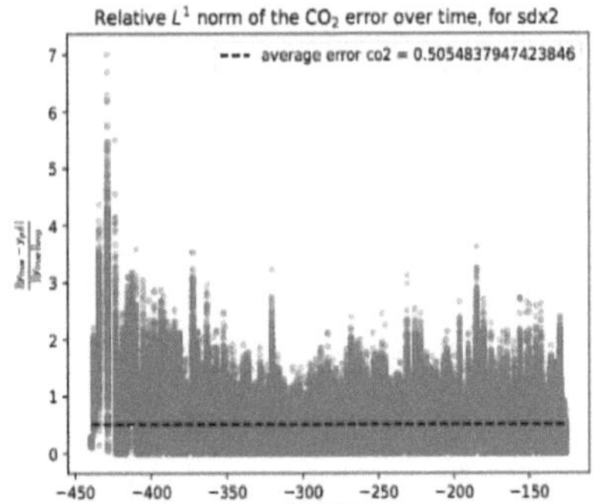

(b) The CO_2 relative error over time.

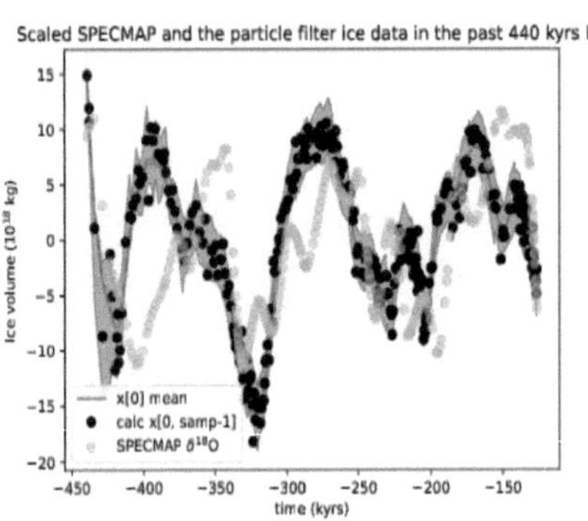

(c) The ice data over time.

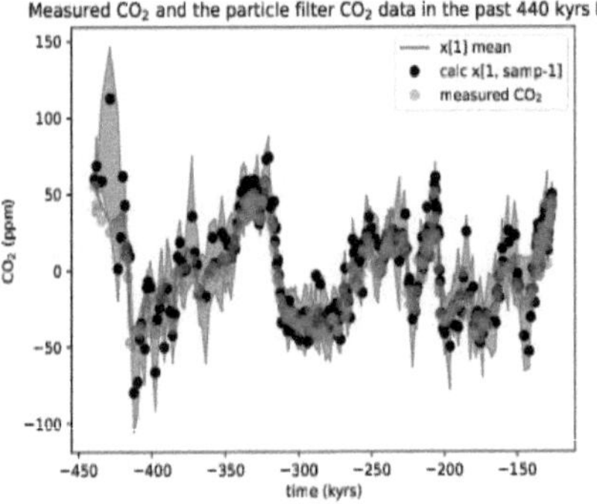

(d) The CO_2 data over time.

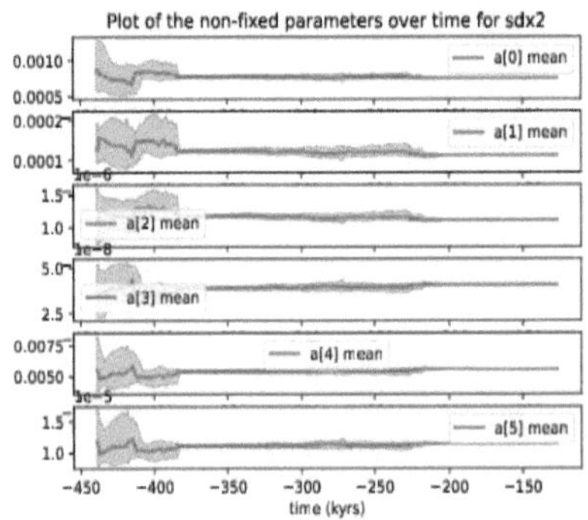

(e) The parameter distribution over time.

Figure 3.36: Results for the SM90 model without insolation forcing.

Then, for the BR22 model, here are the results for the insolation forcing case:

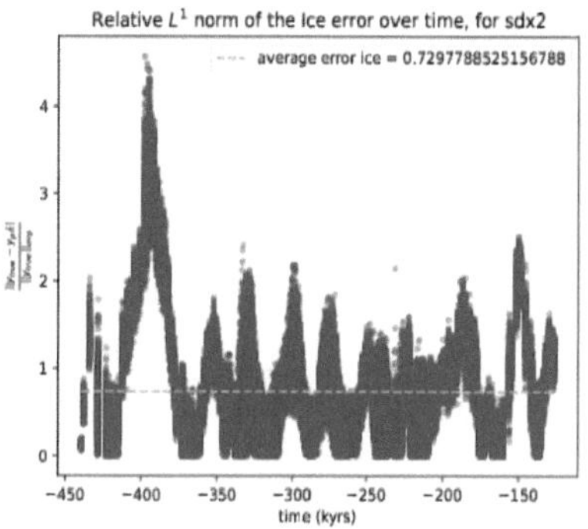

(a) The ice relative error over time.

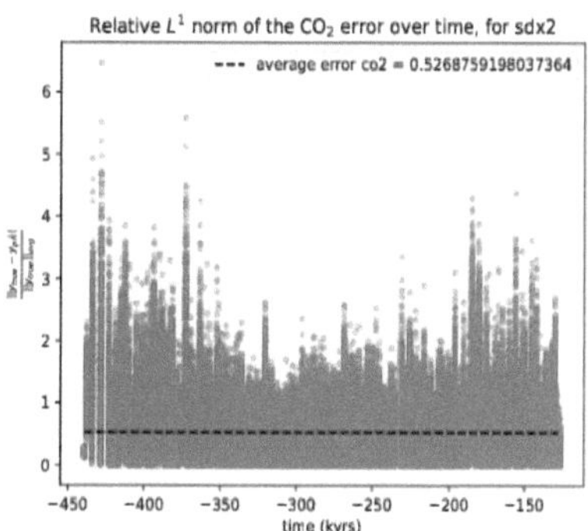

(b) The CO_2 relative error over time.

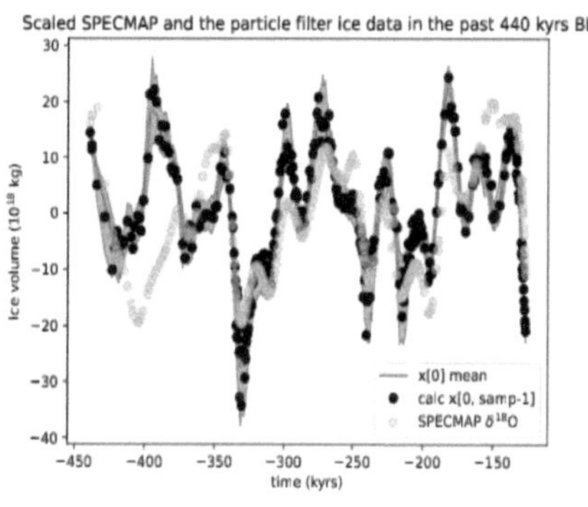

(c) The ice data over time.

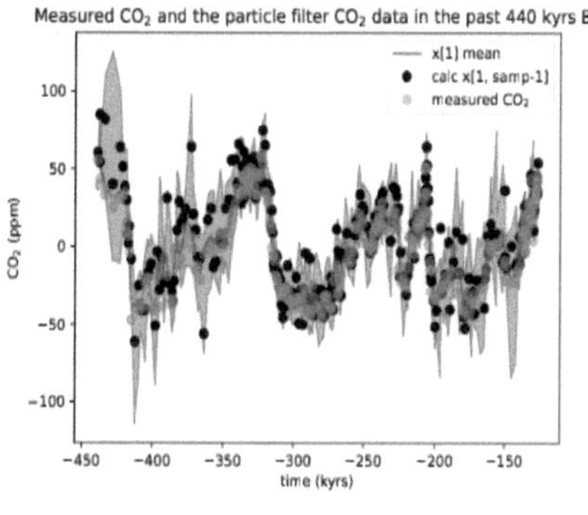

(d) The CO_2 data over time.

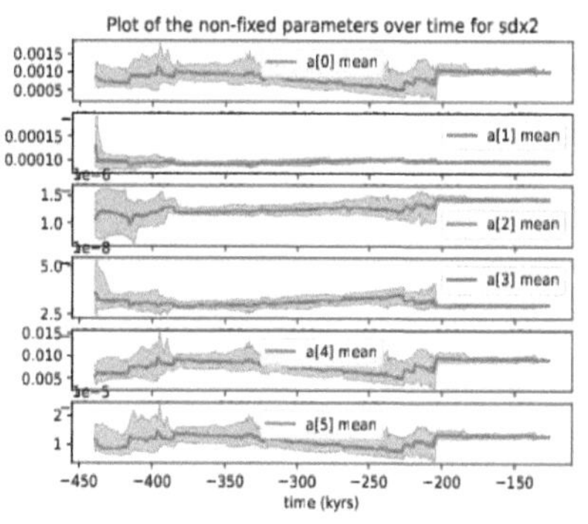

(e) The parameter distribution over time.

Figure 3.37: Results for the BR22 model with insolation forcing.

and the results for the BR22 model without insolation forcing:

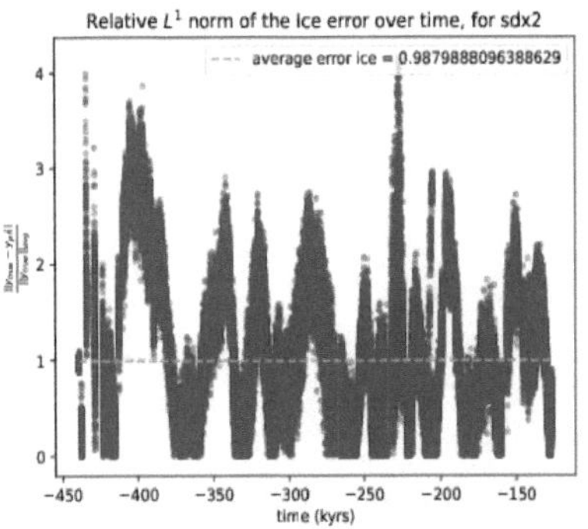

(a) The ice relative error over time.

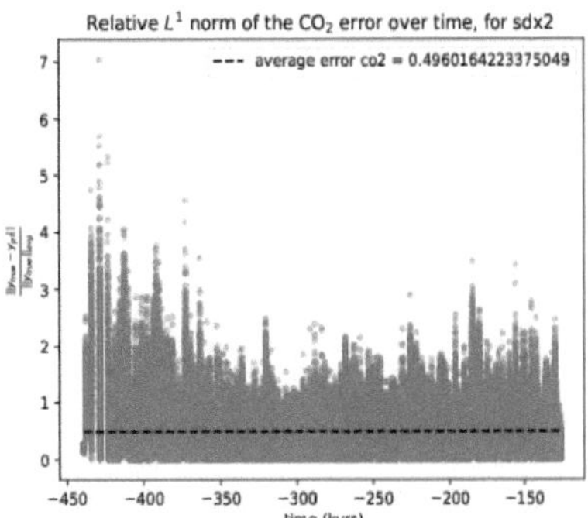

(b) The CO_2 relative error over time.

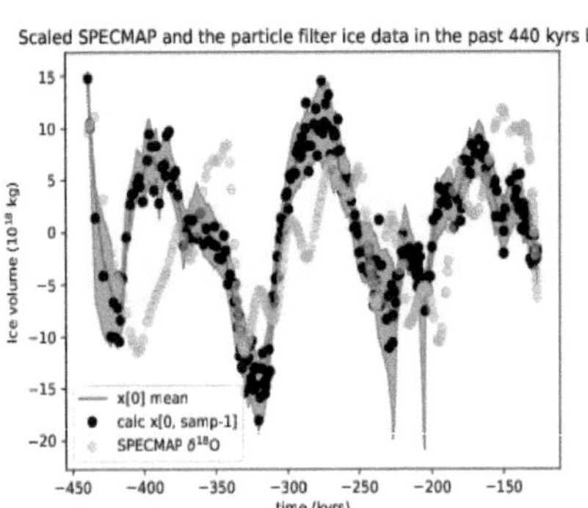

(c) The ice data over time.

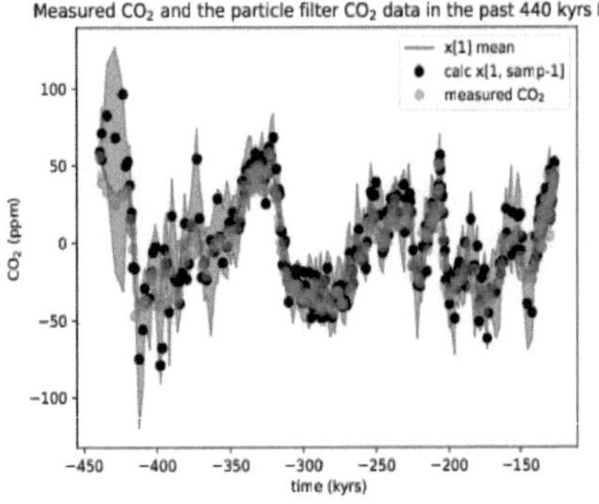

(d) The CO_2 data over time.

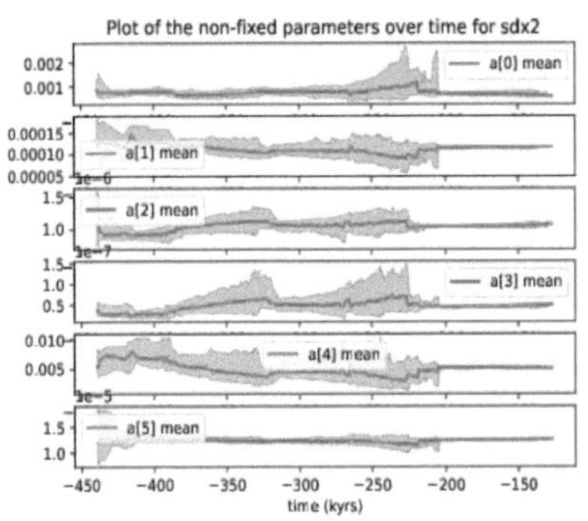

(e) The parameter distribution over time.

Figure 3.38: Results for the BR22 model without insolation forcing.

Finally, for the unforced SS87 model, here are the results:

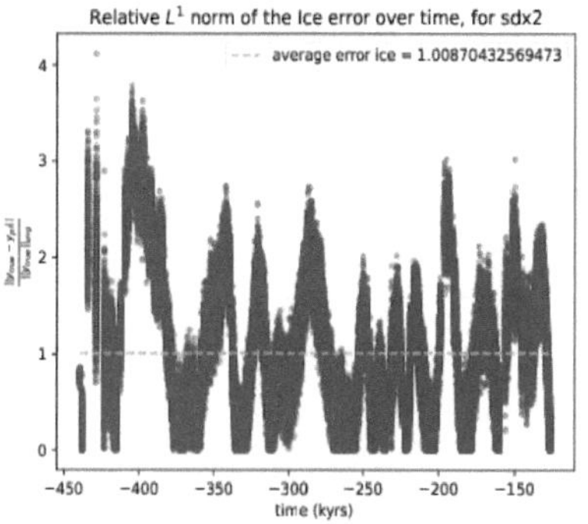

(a) The ice relative error over time.

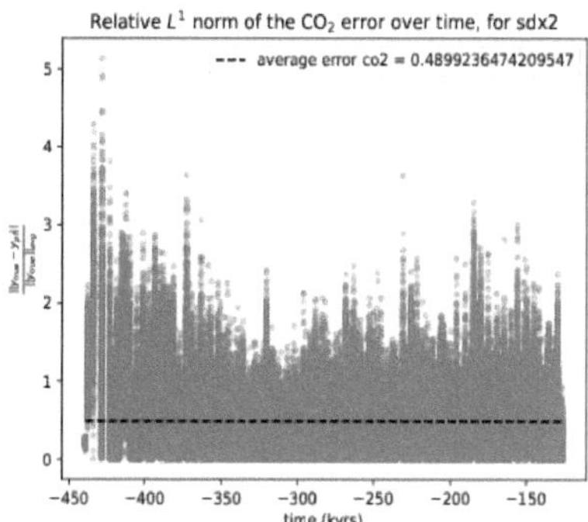

(b) The CO_2 relative error over time.

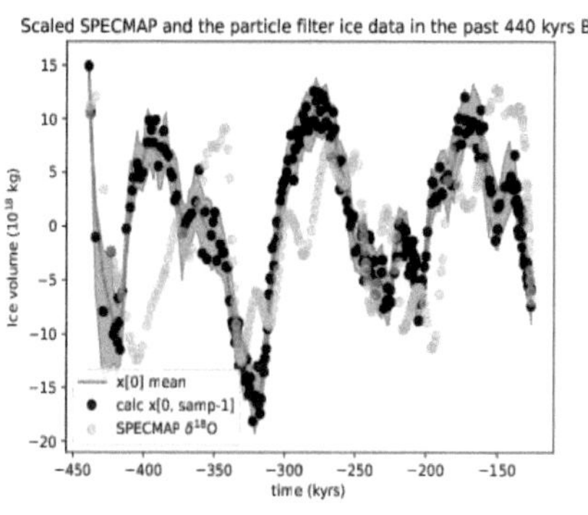

(c) The ice data over time.

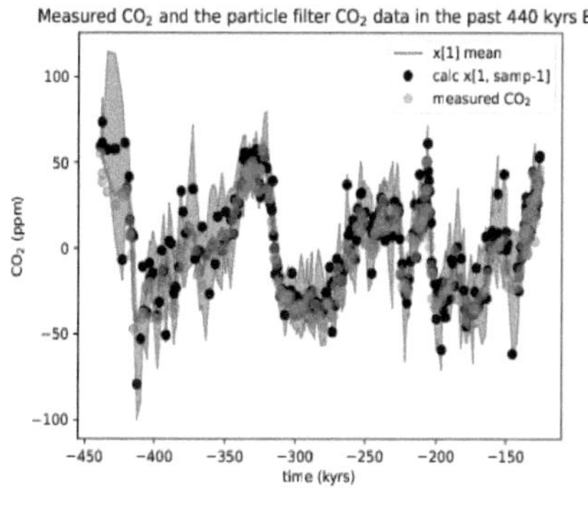

(d) The CO_2 data over time.

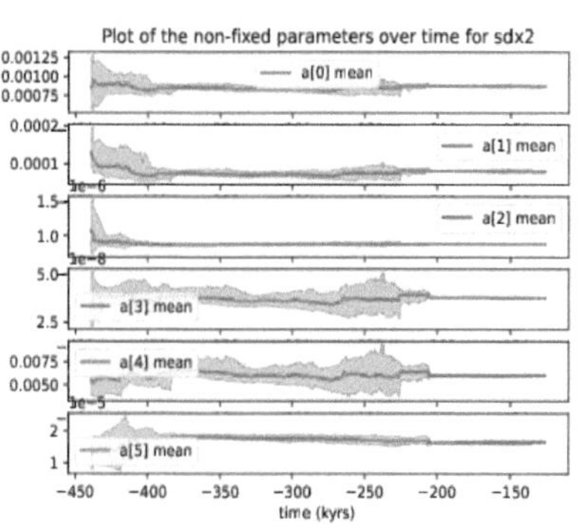

(e) The parameter distribution over time.

Figure 3.39: Results for the SS87 model without insolation forcing.

Now, we would like to present results of the error bars of the measure E for each component (CO_2, ice volume) of each model which will give us another perspective on how well the particle filter did on each model.

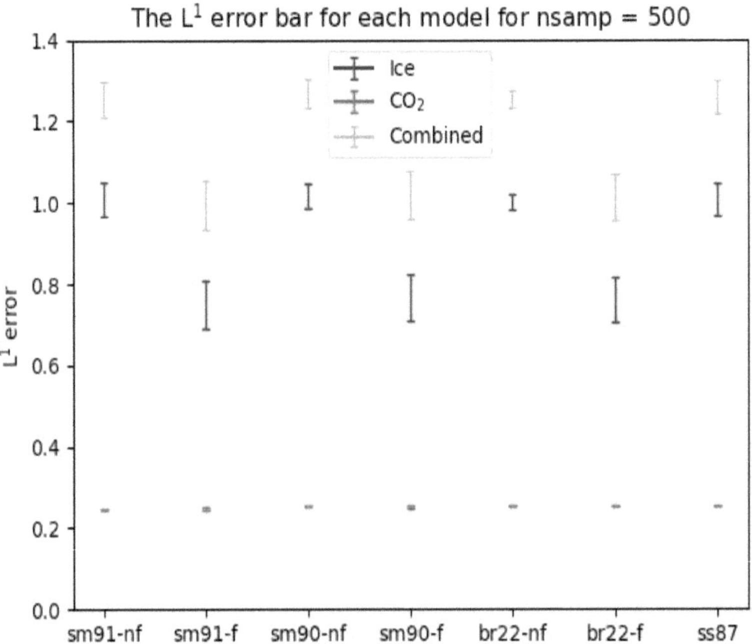

Figure 3.40: The error bars for the measure E of each model for nsamp = 500.
Note: "nf" = "no forcing" and "f" = "forcing"

A couple interesting features of Fig. 3.40 should be noted: 1. the different error levels for the different models is essentially all due to differences in the error in ice volume alone, 2. each model with no external forcing all have similar error levels (similarly for the forcing case), 3. the error levels for the forcing case, regardless of the model, is always lower than the error bars for the non-forcing case.

The error magnitudes observed in Fig. 3.40 can be explained using the figures for the ice data over time (subfigure (c) in Figs 3.33-3.39) as the particle filter predictions miss the peaks between -350 and -300 kyrs for each of the models without forcing, whereas models with forcing all do better within that interval regardless of the model chosen. Another interesting feature in the ice volume results is, regardless of what model, the particle filter seems to miss the valley at around -400 kyrs. It is also noted that from -200 kyrs and on, the parameters all seem to "settle" (subfigure (e) in Figs 3.33-3.39). This convergence of the parameter distributions can be explained from looking at the CO_2 data graph over time (subfigure (d) in Figs 3.33-3.39), where the particle filter does seem to be "locked in" to the measured CO_2 data, i.e. follows the measured data very well.

We return now to the question of model selection. In this Section we have used the particle filter to fit model parameters to CO_2 data and then measured the model performance with respect to CO_2 data and ice volume proxy data. Since the ice volume data was not used in the particle filter, it represents a more rigorous independent test of the models. As shown in Fig. 3.40, the main result of the ice data on model selection is to affirm that insolation forcing is a critical feature of the model. Beyond the necessity of insolation forcing, there was a slight preference for the forced SM91 model as giving the lowest overall error, but all the forced models had similar overall errors with error bars overlapping. Thus the particle filter does appear to be "selecting" models that use external forcing, but it doesn't seem to discriminate between the forced models based on their long-term dynamics as suggested in [3].

3.5 Conclusion

Following the idea in Crucifix & Rougier [4], we have implemented the particle filter to determine the states and parameters for dynamical systems models of ice ages. We tested our numerical method for the forced stochastic dynamical system by comparing directly to the figures in Crucifix & Rougier, and conducted benchmark tests of the particle filter to determine its performance on both synthetic and actual CO_2 data. What is intriguing is that regardless of the model we tried, sdx2 always gave better and more stable results than running the particle filter using sdx1. In addition, it also appears that a measurement error of sdy $= 5$ ppmv (smaller than 20 ppmv used in Crucifix & Rougier (2009), p. 25), always gave better results as well, regardless of the model used; since sdy is the measurement error for the stochastic model, this result likely has to do with the underlying variability in the time record for the CO_2 data.

We note that the CO_2 data used in Figure 8, p. 26, of Crucifix & Rougier (2009), is somewhat different from us in that the "the Vostok time-scale, called GT4, uses a conventional date of isotopic stage 5.4 of 110 kyr BP" (Crucifix & Rougier (2009), p. 10) whereas we are simply using the one block of data from 0 to 440 kyrs BP from [14] which may in fact eliminate the GT4 bias that was mentioned in Crucifix & Rougier (2009). Despite some differences in the CO_2 data, the results for the SM91 model, in particular, was consistent with what was found in Crucifix & Rougier (2009) with sdx $=$ sdx2 and sdy $= 5$ ppmv in that the particle filter tracked the CO_2 well.

The research here suggests that more work is needed to evaluate dynamical systems models of ice ages. Part of the issue is the data. By nature, the data itself a proxy based on geochemical measurements in which the measurement uncertainty includes the geological time stamp. Some data, such as the SPECMAP data $\delta^{18}O$ depends on more than one system variable (eg. ice volume and temperature).

With regard to the particle filter, the Liu & West algorithm is a powerful tool for using data to determine model parameters. In our case it could determine six parameters of a three state model using measurements of one of the states. In our work we made choices for quantities such as the process noise and the measurement error based on benchmarking the algorithm. A potential future project might be to incorporate these parameters as part of the parameter set determined by the particle filter.

Chapter 4

Summary

We developed a particle-filter code to filter for states and parameters and implemented it for different dynamical-system models for ice ages with the aim of determining if the particle filter could select models based on the available CO_2 data. This modeling approach was put forth in a 2009 paper by Crucifix and Rougier [4] for a single ice age model. In a subsequent paper [3], Crucifix alludes to unpublished results that suggest that the particle filter was not able to distinguish been different models based on long term dynamics. Our motivation for this work is to provide a detailed investigation into the issue of ice age model selection using the particle filter.

The ice age models we considered are a family of three-state, first-order stochastic dynamical systems for global values of CO_2, glacial ice volume and ocean temperature. In addition to model parameters, the effect of variations in solar energy input from the sun due to long term variations of the Earth's orbit was included as a deterministic forcing term that could be toggled or toggled off. Some of the model parameters were taken as fixed, while others (typically those associated with nonlinear terms) were taken as unknown and to be determined by the particle filter. The size of the stochastic forcing terms Σ was also unknown, and we considered two cases which we called sdx1 (small) and sdx2 (large).

The particle filter method is based on using a large number of discrete samples ("particles") to generate a forward-in-time Bayesian approximation to the probability distributions of the model parameters and model states. As the model and parameter distributions are evolved in time, each data observation is used to select particles which are good predictors of the data and reject particles which are bad predictors of the data. The Liu and West algorithm [12] also includes a dispersion term so that the cloud of particles remains dispersed and able to continue to explore parameter space. Computational parameters in the algorithm are the number of samples, the estimated size of measurement error (sdy), and a resampling threshold.

In Section 3.1 we validated our numerical model and conducted benchmarking tests of the particle filter on synthetic CO_2 data generated by the numerical model. In Sections 3.2 and 3.3 we applied the particle filter to the actual CO_2 data on various ice age models. Finally, in Section 3.4 we used the particle filter to fit the actual CO_2 data and measured model performance with respect to CO_2 and an ice volume proxy.

The particle filter and other numerical methods were first validated using "synthetic" data, i.e. data that was generated from the SM91 model itself. This was done by checking our results against Crucifix & Rougier (2009) (e.g. verifying the first two figures in Figure 7, p. 23, in Crucifix & Rougier (2009)), as well as confirming that the particle filter will recover synthetic data. Convergence to the data was observed from the figures and in terms of the error measure E as defined in Eq. (3.2). It was found that having a process error Σ (in this paper it was called "sdx") larger than what was listed in Table 1 of Crucifix & Rougier

(2009) and a measurement noise sdy (M. Crucifix calls "observation error") to be smaller than was on p. 25 of Crucifix & Rougier (2009) tend to have a "lower" E profile in terms of the graphs in Section 3.1.

Next, we moved on to applying the code to real data in Section 3.2. Interestingly, it was also found that having an sdx larger than what was listed in Table 1 of Crucifix & Rougier (2009) and an sdy smaller than was on p. 25 of Crucifix & Rougier (2009) led to very good results in terms of the measure E. Overall, our results were consistent with what was found in Crucifix & Rougier (2009). In fact the central parameters listed in Table 1 of Crucifix & Rougier (2009) were, on average, within one standard deviation from the parameters obtained in our results.

Next, a sensitivity study was done using the particle filter to see how well it did in terms of selecting the "best" candidate for an Ice Age model. To evaluate each model we calculated an error measure E, and determined E based on multiple trials to determine error bars for E. Based on this definition, the particle filter did not appear to select the "best" candidates for Ice Age models if one focused on performance on CO_2 data alone.

Finally, we explored how well the models would predict a proxy for ice volume, given only the CO_2 data as input. Here there was a clear indication that insolation forcing was a critical component of the model, but beyond this criterion the models we considered performed similarly. Based on our analysis, we have confirmed the suggestion of Crucifix [3] that the particle filter method may not be able to select an ice age model based on long-term dynamics.